Potentials of PLANT BIOTECHNOLOGY

Potentials of PLANT BIOTECHNOLOGY

Editor
P.B. Bansal

2006
Gene-Tech Books
New Delhi - 110 002

ISBN 81-89729-02-0

Published by : GENE-TECH BOOKS
4762-63/23, Ansari Road, Darya Ganj,
NEW DELHI - 110 002
Phone: 41562849
e-mail: genetechbooks@yahoo.co.in

Printed at : Tarun Offset Printers
Delhi

PRINTED IN INDIA

Preface

Plant biotechnology covers a vast area of methodologies and technologies, ranging from specific tools to improve conventional plant breeding and production, to advanced and powerful tools that breaks with existing norms and principles. The potentials and limitations of plant biotechnology in relation to plant breeding and production are therefore a complex area, and some preliminary information on long term environmental, social, economic and even ethical effects has caused serious concern among governments and non-governmental stakeholders.

The immediate use of biotechnology is in crop production, where the quality and production potential of seeds could be significantly improved, more rapidly and better controlled and through more efficient use of the scarce resources than through use of the traditional plant breeding methods. The quality of seed is only one of many parameters influencing crop production, its quality and size. Others include land tenure/ ownership; availability of inputs such as seed and fertilizer and financial services, and access to markets. The yield potential and value of the planting material are, however, of paramount importance for all the additional investments in crop production. These include labour, mechanisation, fertilizer, irrigation, marketing and capital.

The potential role of modern plant biotechnology for ensuring sustainable food production and alleviating poverty and malnutrition in developing countries is at present under intense global debate. The debate is multifaceted and comprises elements such as the potential of the technology, intellectual property rights, trade, potential risks, ethical issues, as well as more ideological aspects.

This is authentic work on plant biotechnology bring together the principles and practice of contemporary plant biotechnology. It includes the techniques of plant genetic modification—applications of plant biotechnology, crop improvement in agriculture and a production system for pharmaceutical proteins. It also describes ethics, safety issues and legislation within the business of plant biotechnology. It will be an essential reading for students, researchers and professionals in plant biotechnology and related sciences.

Editor

Contents

1

Emerging Trends in Plant Biotechnology

Achieving any of the new applications of plant science will require substantial private and public investments, and a wide range of scientific and communication skills. The required scientific skills lie not only in gene technology, but also in the related fields of plant breeding, agronomy and physiology, food and nutrition and in natural resources management.

The contributions of gene technology to today's agriculture are already substantial. Discoveries in gene technology have led to:

— Better Understanding of how plants function, and how they respond to the environment.

— More targeted selection objectives in breeding programs to improve the performance and productivity of crops, trees, livestock and fish, and post harvest quality of food.

— Use of molecular (DNA) markers for smarter breeding, by enabling early generation selection for key traits, thus reducing the need for extensive field selection.

— Molecular tools for the characterisation, conservation and use of genetic resources.

— Powerful molecular diagnostics, to assist in the improved diagnosis and management of parasites, pests and pathogens.

— Vaccines to protect livestock and fish against lethal diseases.

In crop agriculture, applications of gene technology are making major contributions to present day agriculture by the development of new (conventional) crop varieties, through the use of marker assisted selection. The other important application of gene technology in agriculture is in the development of novel, transgenic plant varieties. Here new genetic instructions are introduced into the crop by laboratory-based molecular methods, leading to new plant varieties that have been genetically modified for a specific trait.

It seems likely that in most countries the applications of gene technology to agriculture will be a two-stage process. Firstly, there are many applications of gene technology that can be used to improve the management and efficiency of present agricultural practices. These include the use of molecular markers in smarter breeding, new diagnostics and vaccines. Secondly, there are options for the targeted introduction of transgenic strains, genetically modified for one or more specific traits. Although transgenic strains of various species of crops, trees, livestock and fish have been developed experimentally, only transgenic crop varieties are in widespread commercial use in agriculture.

Cultivation of Transgenic Crops

Broadly, the first wave of genetically modified crops, which are in commercial use, address production traits; the second wave, which are mainly under development, address quality and nutritional traits; and the third wave for the future address complex stress response traits and novel products able to be produced in plants. The scientific basis of dealing with each of these three groups of traits is increasingly complex.

The production traits targeted in the first wave of transgenic plant varieties specifically addressed the economic and environmental costs of chemical management in large-scale agriculture. An important factor in the initial choice of production traits was the fact that the major early private investors in plant biotechnology were several multinational chemical companies. The long-term viability of chemically based agriculture was being questioned as potentially:

— damaging to human health;

— damaging to the environment, due to chemical damage to living organisms, and excess chemical run off into water courses;

— less effective, due to the build-up of pesticide tolerance in target pests and diseases, thus shortening the economic life of agri-chemicals;

— less feasible due to the difficulty in discovering new agri-chemical compounds.

Biologically based management strategies were been sought, not only to reduce chemical use in agriculture, but .also to find more powerful ways to increase sustainable productivity and improve quality. Strategies for integrated pest management and later integrated crop management were developed in some areas. These strategies aimed to reduce inputs of chemical pesticides, herbicides and fertilisers, maximise the effectiveness of natural enemies for pest control and make judicious use of host-plant resistance for disease and pest control.

In regard to the development of novel genetic approaches for specific production traits, a combination of new scientific possibilities, business opportunities, and decreasing viability of chemically-based agriculture led to the targeting of particular production traits (insect resistance and herbicide tolerance) and their subsequent commercial development into new transgenic crop varieties.

The first transgenic plant was produced experimentally in 1983. The first commercial cultivation was in 1995. By 2001, there were almost 53 million hectares of genetically modified crops growing in 13 countries. The most commercially important of these crops are soybean, corn, cotton and canola (oil seed rape). The traits these new plant varieties contain are mainly insect resistance (corn, cotton), herbicide resistance (corn, soybean), delayed fruit ripening (tomato) and virus resistance (papaya) . Many other crops and traits are under investigation but most have yet to be taken through to practical use.

Gene Production Traits

The developers of the first generation of genetically modified (transgenic) crops faced a number of technical limitations that constrained the choice of crops and traits that have been taken through to full product development. These constraints included:

(1) The availability of genes controlling traits that could be manipulated. Initially, only traits controlled by single genes could be manipulated; and single genes control only a limited number of traits.

(2) The efficiency of the methods to produce genetically modified plants that express the desired trait consistently under field conditions;

(3) The need to meet evolving regulatory requirements for new crop varieties (and other genetically modified

organisms) containing genes from outside their normal range of hybridisation.

Single Gene Traits

Initially, only certain traits that were controlled by a single gene could be genetically manipulated for the development of transgenic crops. Thus, single genes that conveyed resistance to certain species of insects (Lepidoptora) by producing a toxin, were derived from a soil-borne bacterium (Bacillus thuringiensis) and transferred to several plant species. This discovery led ultimately to the commercial production of Bt cotton and Bt corn. Similarly, various single genes that conferred tolerance to selected herbicides were transferred to soybean, canola and corn. In some instances, these two traits have been combined, to produce insect and herbicide tolerant corn and cotton. These applications have led to reductions in the amount of pesticide being used for insect control in cotton and corn.

The careful targeting and correct management of single gene traits is critical for their successful use in agriculture. The management of single gene traits is important so as to avoid the boom/bust cycles typical of single gene resistance when used previously in agriculture. For example, when single gene traits are being manipulated to enhance pest or disease resistance in a plant, there is a risk that the pest or pathogen will evolve so as to overcome this host resistance. This occurs with conventionally bred plant varieties, which are usually replaced after a number of years as the pest or pathogen evolves (eg wheat varieties with rust resistance). Yet, there are also instances where conventionally bred plant varieties with single-gene based disease resistance has been stable over many years, especially for bacterial and virus diseases.

In the deployment of new transgenic varieties of Bt-crops, such as cotton and corn, in broad scale agriculture, much effort has gone into devising and implementing specific crop management arrangements that mitigate against the evolution of resistance in the target pest. These crop cultivation regimes include leaving some of the field as non-transgenic, susceptible crops (refugia for the insects), which reduces the evolutionary pressure on the pest to evolve to overcome the pest resistance in the plant. Another strategy is to include in the plant two or more different genes for pest resistance (eg two different Bt genes). This gene stacking strategy makes it more difficult for the pest to evolve, as it has to overcome two resistance genes with different modes of action.

Complex Traits

Most characteristics of food are controlled by more than one gene. Thus taste, aroma, color, nutritional composition and other aspects of food quality are the result of complex biochemical reactions within the plant before and after harvest. Crop yield is also a complex characteristic, with many genes involved in plant development, flowering, and yield components.

Crop responses to stress during cultivation are often controlled by many genes, which stimulate complex biochemical pathways within individual plant cells. For example, many genes control plant responses to fungal infections. Similarly, some plants have developed means to respond to environmental stresses such as drought by changing their metabolism to accommodate having less water available (eg sorghum).

Emerging scientific developments are enabling complex traits to be addressed, with the intention of developing new products of potential value for agriculture, human health and the environment. These include traits in the following categories:

Increasing sustainable agricultural production, by the cultivation of crops that are better able to tolerate biotic stresses (pests, diseases and weeds) and abiotic stresses (drought, salinity, and temperature stress). Delivering health benefits through more nutritionally beneficial foods, with higher content of essential vitamins and minerals, especially in staple crops such as rice. Reducing allergenic, carcinogenic and/or toxic compounds in certain plants may also be possible, so that they are safer sources of food. (eg reduced cyanide content in cassava; removing allergenic content of nuts; modifying oil content of certain plants to produce more long chain, poly-unsaturated fatty acids).

Using plants for pharmaceutical production: Certain plants may provide a platform (bioreactor) for the more economical and efficient production of specific proteins, such as vaccines against human diseases, and other pharmaceuticals. This approach is being used successfully in microbes and transgenic animals that have been genetically engineered to produce certain high value pharmaceuticals (eg insulin in microbes; blood clotting factor in sheep).

Using plants for production of products for industrial purposes: These compounds may include novel compounds such as biodegradable plastics and industrial strength fibers, as well as the more efficient production of common plant products such as starch and alcohol used for industrial purposes.

Environmental benefits: Using plants (and microbes) to mitigate the effects of industrial pollution (bioremediation), by increasing their ability to remove and/or break down toxic compounds in the soil. Recent demonstrations include transgenic plants changing mercury into a less toxic form; plants with improved abilities of Cd accumulation; and plants degrading industrial waste and harmful substances.

The attractiveness of the new targets amongst complex traits is tempered by the fact that they are technically difficult, requiring the expression and control of multiple genes, often involved in different biochemical pathways. Recent scientific developments confer the ability to study the structure and function of all the genes within an organism simultaneously (through genomics), as well as the protein products they code for (through proteomics). It is also possible to study the role of all the chemical compounds in the metabolism of the cell (through metabolomics). These emerging scientific developments are being greatly assisted by powerful computing and statistical techniques that enable the assembly, interrogation and interpretation of large databases (through bioinformatics). New terms are being coined to describe these rapidly evolving branches of science and the techniques on which they are based.

These new fields of science are using a series of sophisticated techniques to locate and characterise genes, proteins and other compounds, understand their functions and the means to manipulate them for new purposes. Most importantly, several methodologies have been developed that permit the study of genes, proteins and other metabolites collectively rather than individually. The emerging scientific possibilities also pose new challenges in the assessments of the risks and benefits of potential new products to human health and the environment. Some of these potential products are meant for food or feed use, while others are intended for use as pharmaceuticals, and others as compounds for industrial uses. Some will require inter-specific transfer and control of multiple genes. Others will rely on switching on (or off) and better regulating genes that are already present in the organism but not usually expressed.

New scientific developments also offer potential means to overcome some of the risks involved in the cultivation of genetically modified crops. These include limiting the unintentional movement of genes out of the target crop (through gene containment), where such movement may pose a risk to biodiveristy or to the environment. Better food safety assessments of any unintended changes in the composition of foods may be undertaken by assessments of the content of whole foods (through metabolomics).

The challenge is how emerging scientific discoveries, such as those in the rapidly evolving fields of genomics, proteomics and metabolomics, amongst others, can be translated into safe applications of biotechnology that will lead to new varieties of crops, novel foods and new products that deliver benefits for society. These new applications and their risks and benefits will differ in different parts of the world. Careful thought needs to be given to identifying the most suitable targets and desirable traits for future research and development efforts, in different countries and environments.

Platform Technologies

Rapid technical advances are occurring in three major areas: (1) DNA sequencing; (2) genome analysis, and (3) computational biology (bioinformatics). Firstly, developments in DNA sequencing have made the acquisition of whole genome sequences possible. These data, when interpreted with the assistance of bioinformatics, can provide a complete listing of all the genes present in an organism, (its genetic blueprint). The first genome sequence of a higher organism was published in 1996. More than 23 genome sequences are available, and some 60 or more genome sequencing projects of a wide variety of organisms, including plants, animals, parasites, and microbes, are under way.

Secondly, different types of technologies have been developed for genome analysis, which speed up the process. What puts this type of genome analysis into a different league is that, with the immense increase in the amount of DNA sequence data available, it is possible to scan whole genomes rapidly and to develop a systems approach for mapping genetic traits.

There are continuing improvements in molecular techniques so as to reduce their costs and increase their speed and efficiency in dealing with large numbers of genes. Particularly important are cDNA microarray techniques. These are used in functional genomics to identify how each gene responds to a specific environmental stress. Similar techniques are being developed for the study of proteins, through proteins arrays.

For example, in microarray analysis, when all the genes in a plant (about 25,000 in Arabidopsis thaliana) are placed on a glass slide and subjected to a sequence of environmental stresses, it is possible to determine the several genes that are most important in the organism's overall response to stress. These genes will be those that react to all the stresses. This small number of genes can then be studied in more detail for their specific functions and potential use in the development of stress-tolerant plant varieties.

Bioinformatics

It is possible to use the developments in bioinformatics to understand the complex genetic interactions involved in growth, development, and environmental responses. Developments in bioinformatics are allowing the prediction of gene function from gene sequence. Thus from genome sequences of DNA it is possible to build a theoretical framework of the biology of an organism. This forms a powerful base for further experimentation. In addition, as the numbers of physical and genetic maps of different species increase, it becomes possible to compare these across different organisms (through comparative genomics), be they microbes, plants or animals, and to significantly reduce the time required identifying important genes. Some of these genes are conserved (shared) between organisms. These technologies allow novel approaches to addressing biological problems.

Plant Genome Mapping

Much research in plant sciences is done on model plants such as Arabidopsis thaliana, a small crucifer plant belonging to the Brassicaceae. Plant genomes vary greatly in size, ploidy, and chromosome number. Arabidopsis thaliana was chosen as the preferred model plant for genomic and related biological studies because of its small genome and short generation time. The knowledge acquired on model plants has spill over benefits to economically important crops in the development of novel traits.

The first plant genome that has been completely sequenced is Arabidopsis thaliana, as a result of the international Arabidopsis Genome Initiative, which commenced in 1996. The Arabidopsis genome sequencing shows both structural features of the genome and gives information about the function of several genes. The total genome holds about 25,500 genes. About half of these genes appear to be specific to plants.

The genomic sequencing of economically important crops is also being undertaken. The most advanced are the several sequencing projects on rice, to map the indica

(Oryza sativa var indica) and japonica (Oryza sativa var japonica) rice types. Some of these projects are being undertaken by public consortia, led by scientists from China, Japan and the international rice genome sequencing project, for indica rice. Others are being undertaken by private companies, for japonica rice. Almost half the genes so far identified in rice are similar to genes that occur in Arabidopsis thaliana. A maize genome-sequencing project is also in progress. It appears likely that rice, maize and other cereals share a large number of common genes.

Functional genomics

The purpose of functional genomics is to understand the role that a particular gene plays in the life of a plant. Several techniques have been developed to assist in the identification of gene function. These include knock-out techniques (whereby an individual genes is disrupted and the resulting mutant phenotype compared with the wild type to identify any changes in phenotype). This technique is being used in rice to identify the function of all the genes in rice.

A completely sequenced plant genome such as rice, for example, will provide a pool of genetic markers and genes for rice improvement through marker-assisted selection or genetic transformation. To fully exploit the wealth of molecular data it is necessary to understand the specific biological functions encoded by DNA sequence through detailed genetic and phenotypic analyses. Thus unlike genome sequencing (structural genomics), functional genomics requires diversity of scientific expertise as well as genetic resources for evaluation.

In many important food crops, national and international public sector research has a large investment in genetic resources and breeding materials, and a long history of understanding biological function and genotype x environment interactions. These scientific and biological resources will become increasingly important in gaining knowledge about the function of genes and in developing molecular markers to assist the breeding process.

Proteomics

Most cellular functions are carried out by multi-protein complexes. New techniques are enabling these complexes to be unravelled, and the functions of individual proteins understood. Techniques for large-scale protein separation, combined with precise approaches that analyze, identify, and characterize the separated proteins, are enabling researchers to investigate cellular function at the protein level. Proteomics has been enabled by the accumulation of both DNA and protein sequence databases, improvements in mass spectrometry, and the development of computer algorithms for database searching.

New techniques allow the identification and quantification of proteins expressed in a particular tissue or in a specific developmental or environmental condition, such as in response to stress. These techniques include 2-Dimensional -protein gel electrophoresis combined with mass spectrometry. Recent improvements are making the techniques more sensitive so that they can detect small amounts of proteins.

The recent development of protein arrays will be a powerful method to link genomics with proteomics. In this system, fully active proteins are spotted onto membranes to study protein interactions, protein-nucleic acid interactions or protein-ligand interactions.

Metabolomics

Besides the integration of data on protein function and activity, information on metabolite levels in the cell is critical to obtaining a holistic view on a biological process and its functional biocomplexity. Examining changes in metabolic profiles (through metabolomics), can be an important part of an integrative approach for assessing gene function and relationships of phenotypes. Modern high-resolution techniques allow the establishment of a profile of all the metabolites present in a specific plant tissue. By use of improved tools for analytical chemistry a variety of previously unidentified biochemical pathways can now be understood.

Metabolomics can provide information on metabolic network regulation in response to genetic and environmental perturbations, leading to a better understanding of plant responses to stress. Genes encoding the biosynthetic enzymes can also be more easily identified and consequently the production of secondary and intermediary metabolites in crop plants can be envisaged. The study of secondary metabolites is of particular interest to the pharmaceutical industry, since most drugs are based on plant derived products.

By exploring the three levels of genomic analysis (transcriptome, proteome and metabolome), extensive databases of quantitative information are being developed about the degree to which each gene responds to environmental stimuli. These stimuli may come from biotic and abiotic stresses such as pathogens, pests, drought, salt; from chemicals such as phytohormones, growth regulators, herbicides and pesticides; or from changes in developmental processes such as germination and flowering. These databases will provide insights into the set of genes that control complex responses and will create powerful opportunities to assign functional information to genes of otherwise unknown function.

Technologies for Gene Manipulation

The essential tools of gene technology are the techniques

that enable one or more genes that control a particular trait to be transferred within or between species, switched on, and made to express themselves in the right place, at the right time and in the right amount. Thus several enabling technologies are required to facilitate the detection, transfer, and expression of genes. Initially, only a limited number of plant species and sometimes only a small number of strains within a species could be transformed and regenerated.

Much current research is directed at improving the enabling techniques for gene technology, so that the techniques become faster, more precise, less expensive and more widely applicable to all plant species. For example, improved techniques are being developed to enable the introduction and simultaneous expression of multiple genes that control a particular trait. Other techniques being developed are concerned with new ways to manipulate the plant's own genes, by switching them on and off, perhaps in response to particular environmental stimuli.

Producing Specific Compounds in Plants

Metabolic engineering is the in vivo manipulation of biochemistry to produce non-protein products or to alter cellular properties. The products may be native to the plant or novel (expressed after the introduction of genes from another source). The non-protein products able to be produced by plants include:

— alkaloids such as quinine,

— lipids such as long-chain polyunsaturated fatty acids,

— polyterpenes such as rubber,

— structural components such as lignin,

— osmoprotectants such as glycine betaine,

— aroma compounds such as S-linalool in tomatoes,
— pigments such as blue delphinidin in flowers,
— vitamins such as folic acid,
— biodegradable plastics such as polyhydroxyalkanoates.

Recent research shows that the following applications of metabolic engineering are technically possible in plants at the experimental level: Increasing vitamin A content; increasing essential oil production (Mahmoud and Croteau, 2001); decreasing lignin deposition; stimulating the bioconversion of secondary metabolites to medicinally important alkaloids; improving tomato flavor and producing biodegradable plastics in plants. Several of these products are now in development phase and are likely to be coming forward for regulatory approval over the next several years.

Regulatory Requirements

The revolutionary nature of the discoveries in gene technology has also raised concerns as to the safety of genetically modified foods for human consumption and the potential impact of genetically modified crops and other living modified organisms (LMOs) on the environment. These concerns arise largely because the first generation of genetically modified crops have been produced by the introduction into plants of genes from other phyla, with whom the plants would not normally cross in nature.

Regulatory/risk issues that are being addressed through new research developments include:

(1) Limiting gene flow to related and wild species, by gene containment;

(2) Control of trait expression and movement, to minimise impact on non-target species; for example, by limiting

transgene expression to specific plant parts and times in the life cycle;

(3) Removal of DNA from selectable markers and promoters, especially when these are derived from bacteria or viruses.

Selectable markers are used to demonstrate that a transformation event has occurred and to select these transformed individual plants in the laboratory. These selectable markers have often been genes for antibiotic resistance derived from bacteria. The use of antibiotic resistance markers especially has raised some concerns that these may contribute to the further development of antibiotic resistance in humans, by horizontal gene transfer of DNA from genetically modified foods and crops to humans and animal pathogens. The likelihood of this happening is remote. Nevertheless, new selectable markers such as those conveying a green fluorescent pigment are replacing these antibiotic resistant selectable makers. Another approach is to remove the selectable marker after the transformed plants have been selected in the laboratory but before they move into field testing and product development.

There are also some additional regulatory issues emerging from the new applications of plant biotechnology. These issues include:

— Ensuring that when plants are used to produce products for industrial uses, these products do not inadvertently enter the food chain;

— Inter-action with pharmaceutical regulations, where plants are used to produce vaccines and other medicinal products;

— Identifying the extent of regulation required for nutritionally modified foods (nutriceuticals), to ensure that these foods meet their makers' claims for improving nutritional quality, for example by the expression of nutritionally significant levels of essential vitamins and minerals;

— Identifying where crops with modifications to genes normally present in the crop species may need to be regulated differently from those containing transgenes from other phyla;

— Identifying where crops with complex traits regulated by several genes may have a reduced liklihood of transferring these complex traits to related species and wild relatives, than transgenic crops carrying single gene traits, and any differences in regulation and safety assessments that may be required.

— Regulatory systems need to be sufficiently flexible to be able to respond to emerging scientific developments, both in terms of encouraging their use to address risks associated with current applications of gene technology, and in recognizing any new issues likely to arise when new scientific opportunities are applied to agriculture.

2

Plant Genetic Engineering

Genetic engineering of plants is much easier than that of animals. There are several reasons for this:

1) there is a natural transformation system for plants (the bacterium Agrobacterium tumefaciens),

2) plant tissue can redifferentiate (a transformed piece of leaf may be regenerated to a whole plant), and

3) plant transformation and regeneration are relatively easy for a variety of plants.

The soil bacterium Agrobacterium tumefaciens ("tumefaciens" meaning tumor-making) can infect wounded plant tissue, transferring a large plasmid, the Ti plasmid, to the plant cell. Part of the Ti (tumor-inducing) plasmid apparently randomly integrates into the chromosome of the plant. The integrated part of the plasmid contains genes for the synthesis of (1) food for the bacterium, and (2) plant hormones. Genes from the Ti plasmid that are integrated in the plant chromosome are expressed at high levels in the plant. Overproduction of the plant hormones leads to continuous growth of the transformed cells, causing plant tumors. Rapid, cancerous growth of the transformed plant tissue obviously is advantageous to the bacterium: more food gets produced.

The Ti plasmid has been genetically modified ("disarmed") by deleting the genes involved in the production of bacterial food and of plant hormones, and inserting a gene that can be used as a selectable marker. Selectable marker genes generally are coding for proteins involved in breakdown of antibiotics, such as kanamycin. Any gene of interest can be inserted into the Ti plasmid as well. In principle, one can thus transform any plant tissue, and select transformants by screening for antibiotic resistance. However, unfortunately, there are some complications: (1) it has proven difficult to transform monocots (grasses, maize) by Agrobacterium, and (2) regeneration of plants from tissue culture or leaf discs is not always possible.

A number of genetically engineered plant varieties have been developed. Traits that have been introduced by transformation include herbicide resistance, increased virus tolerance, or decreased sensitivity to insect or pathogen attack. Traditionally, most of such genetically engineered plants were tobacco, petunia, or similar species with a relatively limited agricultural application. However, during the past decade it now has become possible to transform major staples such as corn and rice and to regenerate them to a fertile plant. Increasingly, the transformation procedures used do not depend on Agrobacterium tumefaciens.

Instead, DNA can be delivered into the cells by small, µm-sized tungsten or gold bullets coated with the DNA. The bullets are fired from a device that works similar to a shotgun. The modernised device uses a sudden change in pressure of He gas to propel the particles, but the principle of "shooting" the DNA into the cell remains the same. This DNA-delivery device is nicknamed "gene gun", and has been shown to work for DNA delivery into chloroplasts as well.

Over the last several years, use of the "gene gun" has become a very common method to transform plants, and has been shown to be applicable to virtually all species investigated. For example, transformation of rice by this method is now routine. This is a very important development as rice is the most important crop in the world in terms of the number of people critically dependent on it for a major part of their diet.

Another method to get foreign genes into cereals is by electroporation: a jolt of electricity is used to puncture self-repairing holes in protoplasts (i.e., the cell without the cell wall), and DNA can get in through these holes. However, it is often very difficult to regenerate fertile plants from protoplasts of cereals. Nonetheless, significant advances in overcoming these practical difficulties have been made over the years.

Now even transgenic trees have been created: for example, the gene for a coat protein of the plum pox virus has been introduced into apricot. The plum pox virus leads to the feared Sharka disease, for which there is no cure. The resulting transgenic tree shows a markedly decreased sensitivity to this virus. The reason why continuous exposure of the tree to the viral coat protein leads to tolerance against virus infection is not yet understood, however.

Thus, now there are a number of different techniques to introduce foreign genes into plants. Essentially all major crop plants can be (and have been or are being) genetically engineered, the procedures are now routine and the frequency of success is very high. Even though genetically engineered crops are more costly than the usual ones, they have been rather readily accepted by US farmers provided that tangible benefits can be demonstrated. However, it is questionable whether the farmer in poorer countries can come up with the funds to "try out" and use the new crops.

Another issue in this respect is how genetically engineered crops are perceived by the consumer. Even though in the US there is little resistance to such crops as long as the products can be shown to be safe and advantageous, in other countries (for example in sections of Europe) genetically modified foods are received poorly by the consumer. It is unlikely that there is a rationally sound basis for this rather hostile reaction of the consumer, as most of the crops are the result of human manipulation (such as centuries of breeding) and may have been treated with harmful herbicides and pesticides. Time and education will need to be invested to provide consumers and consumer advocates with a balanced opinion on the acceptability of the origin of their foods.

One area of particular concern for some is the lack of labeling of genetically engineered foods. On the other hand, as so many plants (soybean, corn, etc.) are genetically modified and the nature of the genetic modification is not necessarily easy to explain, it may be simpler to label those foods that are guaranteed free of "genetically modified organisms" or their products. However, it is good to keep in mind that essentially all agricultural products have been genetically modified by traditional breeding, so it may be difficult to define what is actually free of genetically modified organisms. Several plant biotechnology companies have increased their efforts to provide information regarding the full, global scope of impacts of plant biotechnology.

Pharming and Plantibodies

An increasingly viable option is the production of highly valuable enzymes by plants and animals. In addition to production of human proteins in these organisms, other valuable proteins that are currently produced by microorganisms could very well be produced by higher organisms instead. Animal and plant "bioreactors" in some respects may be superior to recombinant bacterial systems, because eukaryotes deliver proteins in the generally more desirable glycosylated form. These organismal systems may also be cheaper than cell fermentation techniques.

Two examples of production of human proteins in plants include the production of human serum albumin in transgenic tobacco and potato, and production of human insulin by tobacco. In both cases, the produced protein appears to be fully effective in humans. Unfortunately, however, one cannot raise his/her insulin level by eating transgenic tobacco leaves, as the protein in most cases will be broken down to amino acids before it reaches the blood stream. Therefore, in these cases one cannot escape the practice of protein isolation and purification before transgenic leaves are converted into drugs.

Over the last few years, also antibodies have been produced in plants. Initially, the antibody's light and heavy chains were produced in different plants. But a subsequent cross of these two varieties resulted in progeny carrying assembled and functional antibodies.

Reversible Male Sterility in Plants

Heterozygous individuals often are healthier and stronger than homozygous ones. The only way to guarantee heterozygicity in plants is to make sure self-pollination cannot occur. For most crop plants it was very tedious or practically impossible to exclude selfing. To exclude self-pollination, it would be good to introduce male sterility in plants: progeny from such plants are then expected to be 100% heterozygous (assuming they were pollinated with pollen from an unrelated variety).

To introduce male sterility, a promoter was identified that was turned on exclusively in tapetum cells (a tissue around the pollen sac that is essential for pollen production). This promoter then was linked up to a gene coding for a bacterial ribonuclease (named barnase). This ribonuclease selectively chops up ribonucleic acids. The promoter/ribonuclease construct was then introduced into plants (canola, tobacco, you name it). Because the promoter allows expression only in tapetum cells, the gene construct disrupts only development of the tapetal tissue and its end product, pollen. Plants transformed with this construct were male-sterile but otherwise normal.

Although male-sterile plants are valuable for hybrid seed production, they have limited value when it comes to crop production. Fertility must be restored to crops such as wheat, rice, and tomato, in which the seed or fruit is the harvested product. Fortunately, the ribonuclease is inhibited very much by a simple protein, named barstar. One can thus cross the male-sterile plant with a male-fertile variety in which the gene for barstar has been introduced, and the result is progeny with viable pollen and restored fertility.

Antisense RNA

Antisense RNA refers to nucleotide strands that are produced in a cell and that are complementary to a particular mRNA. Antisense RNA can be produced, for example, by inverting the coding region of a gene with respect to its promoter. The antisense RNA can hybridize with its corresponding mRNA, making it double-stranded. The double-stranded mRNA no longer can be recognized by the protein-synthesizing machinery (the ribosomes), and thus expression of this mRNA is suppressed. Also, in many systems double-stranded mRNA is very unstable and is broken down quickly.

Thus, one can inactivate specific genes while not interfering with others. This is very useful for both agriculture and medicine, and the first examples of practical applications of this RNAi (RNA interference) technology are appearing. As with any new technology, the initial pilot projects are sort of pedantic (including manipulation of flower color and of the speed of fruit ripening). However, more exciting application possibilities abound. In the mean time, antisense technology is contributing to the birth of a new field, reverse genetics. With this method one can turn off genes, and see what the consequences are.

Antisense approaches already are used to protect plants from damage by plant viruses. For example, reversal of a gene from bean yellow mosaic virus (BYMV), and putting it into tobacco under a reasonably strong promoter, has led to a tobacco variety that is quite resistant to BYMV. A similar approach is used to transfer viral resistance to other plants. This finding is of significance, in that currently no effective, environmentally friendly methods exist to control many plant viruses.

Potentials of Genetic Engineering

Perhaps indicative of the large potential and relative ease of genetic engineering, developing countries (particularly China) are progressing rapidly in development and application of genetically engineered crops. Some have gone into commercial production well ahead of similar crops in the US. In China, for example, tomatoes that have been engineered for improved virus resistance have been on the market since late 1992. There are two main reasons for the more rapid commercialization of bioengineered crops in the developing world:

(1) less tight governmental approval mechanisms, and

(2) hungrier populations.

While in developed countries the main value of biotechnological applications may be to reduce production costs, in the developing world a main factor is the production of more food. Indeed, genetic engineering applications seem to be pretty successful to cut down on pathogen-induced losses. For example, genetic modification of papaya plants (expression of the ringspot virus coat protein in the plant) protects very well against the very destructive ringspot virus.

Such "plant biotechnology" or "genetic engineering" techniques are enabling scientists to pursue exciting novel strategies for plant improvement, most of which would be impossible using traditional plant breeding methodologies. These new lines of investigation have great potential to benefit farmers, consumers and societies worldwide. Indeed, first generation transgenic herbicide and pest resistant crops have already made it to the field. These engineered plants not only promise increased crop yields and better quality products, but also allow the use of chemicals that are more environmentally friendly and pose less hazards to human and animal health. Future generation transgenic crops will not only provide further advances in these areas, but in addition, will lead to the creation of foods that are nutritionally-enhanced to improve human health.

Through the power of genetic engineering, it should also be possible to dramatically increase the use of plants to produce "industrial feedstocks" such as specialty oils for lubricants, precursors of plastics and valuable health-related biomolecules. Such developments would not only directly benefit the consumer, but would also afford farmers greater opportunities in choosing what crops to grow.

Most of the plant genetic engineering projects that are in progress fall under one of the following three general headings, each of which would include "value-added" traits:

— Engineering for Improved Crop Production and Quality

— Engineering for Improved Health

— Engineering for Alternative Nonfood Uses

Some examples of efforts in these areas include:

Genetic Engineering for Improved Crop Production and Quality

— Herbicide Resistance (e.g.: glyphosate resistance)

— Pest Resistance (e.g.: BT toxin for insect resistance)

— Disease Resistance (e.g.: increase resistance to viral, fungal and bacterial diseases)

— Environmental Stress Tolerance (e.g.: increase freezing and drought tolerance)

— Modification of Post-Harvest Characteristics (e.g.: prevent potato sweetening)

— Altered Cell Wall Composition (e.g.: decrease lignin content)

Genetic Engineering for Improved Health

— Edible Vaccines (e.g.: prevent infectious diseases)

— Healthier Oil Profiles (e.g.: increased unsaturated fats)

— Increase Vitamin and Mineral Contents (e.g.: increase iron and vitamins A and E)

— Decrease Food Allergies (e.g.: eliminate allergens in milk, nuts and cereals)

— Improve Protein Content (e.g. high lysine corn)

Genetic Engineering for Alternative Nonfood Uses

— Specialty Oils (e.g.: jet engine lubricants)

— Plastics (e.g.: biodegradable thermoplastics)
— Pharmaceuticals (e.g.: anti-cancer drugs like taxol)
— Hormones (e.g.: insulin and growth hormone)
— "Plantibodies" (e.g.: human antibodies for treatment of infectious and autoimmune disease)
— Vaccines (e.g.: against infectious diseases like cholera and rabies)
— Phytoremediation (e.g.: remove heavy metals and toxic organics)

Clearly, genetic engineering offers a wide range of exciting possibilities for future crop improvement that were not only unthinkable, but actually impossible, before the advent of this powerful technology.

Herbicide Resistance

Weeds, if uncontrolled, have major negative effects on plant yield. Thus, farmers apply a variety of chemicals to prevent weed growth. Unfortunately, many of these chemicals are hazardous to human and animal health. In addition, many are environmentally unfriendly as they persist in the soil and can accumulate in ground water.

This overall situation has lead investigators to seek chemical herbicides that are less toxic and more environmentally friendly. Indeed, a number of such chemicals have been discovered such as glyphosate, commonly known as Roundup (produced by Monsanto). The chemical is quickly degraded in the soil, has low toxicity to humans and animals, and is very effective in killing plants. The initial problem, however, was that the chemical effectively kills a wide range of plants; it would kill the target weeds, but in addition, would kill the crop the farmer was trying to grow. Through the power of genetic engineering, however, genes have been isolated, modified and transformed back into crop species to make them resistant to the herbicide. This, then has made it possible for farmers to use the herbicide to kill weeds, but grow healthy crops.

Improved Stress Tolerance

Environmental stresses including extremes in temperature and drought have a major impact on crop production. It has been estimated that in the U.S., the average annual yield of the major row crops is only 20% of their genetic potential, with most of "missing" 80% being due to environmental stresses. In addition, environmental stresses greatly limit the locations where crops can be grown. For instance, due to freezing temperatures, winter canola cannot be grown throughout the northern U.S. nor throughout most of Canada.

Likewise, the forest industry would dearly like to grow eucalyptus throughout the southeast for paper production, but can't due to periodic springtime frosts. Last January in California, the citrus industry experienced some $600 M in losses due to a couple of days of freezing temperatures. And we are all aware of the losses caused by drought this summer in parts of the Eastern U.S.

Given the importance of stress tolerance, one might think that plant breeding programs would include efforts to increase environmental stress tolerance. Indeed, this is the case. However, because of the physiological and genetic complexities involved in enhancing stress tolerance, traditional breeding efforts have met with little success. The most freezing tolerant wheat varieties today, for instance, are only marginally better than those developed in the early part of this century.

Increase Vitamin and Mineral Content

Vitamin A deficiency affects some 400 million people

worldwide, leaving these individuals vulnerable to infections and blindness. Similarly, iron deficiency affects some 3.7 billion people, particularly women, leaving them weakened by anemia and susceptible to complications during childbirth.

Edible Vaccines

The development of edible plant vaccines has the potential to provide more convenient, less costly immunization strategies and means for implementing universal vaccination programs worldwide. Indeed, vaccination programs remain problematic in many parts of the world, particularly in developing nations. As noted by Dr. Charles Arntzen, "the dramatic impact of modern vaccines is not reaching the developing world where it is most needed." This is due in large part to a lack of equipment needed for making, storing and delivering vaccines, but in addition, includes cultural differences that impede acceptance of injection-based immunization.

However, what if people could eat foods that are part of their normal diet, but are enhanced with subunit vaccines that could immunize against diseases? As Dr. Arntzen wondered a number of years ago, what if bananas could be used to produce edible vaccines? There would be fewer problems with storage of the vaccine; the vaccine could be produced more economically compared to traditional vaccine production methods; and such a vaccine could potentially get around technical and cultural problems associated with injection-based immunization.

Such considerations have led a number of investigators to pursue the development of edible plant vaccines. Ones that Dr. Arntzen is particularly interested in are those that would protect individuals from enteric diseases including cholera and diarrhea, leading causes of infant deaths in the developing world. In his initial studies,

Dr. Arntzen and colleagues produced transgenic potato plants (banana transformation requires further development) that were demonstrated to be effective in immunizing mice against bacteria that cause diarrhea. The potatoes were then used in the first-ever human clinical trails utilizing a genetically engineered food to deliver a pharmaceutical. And they proved successful! Further human clinical tests are now in planning.

Environmental Issues of Genetically Engineered Plants

Biological Pollution

Genetically engineered (GE) plants are created by splicing foreign genetic material into plant genomes, creating new organisms that could never arise in nature. The most common form of gene-splicing is quite similar to the process of viral infection. The gene is first cobbled together with pieces of unstable DNA from disease-causing viruses and bacteria, which are required to "infect" the target plant with the desired gene, overcome the plant's defense mechanisms against foreign DNA, and force the plant to express the inserted gene. Once inserted, both the gene and its viral helpers are passed on to future generations; they may mutate, recombine with viruses, or be transferred "horizontally" to other organisms, with unknown and largely unstudied consequences. While chemical toxins may be cleaned up, genetic engineering truly gives pollution a life of its own.

Chemical Dependence

Almost two-thirds of genetically engineered crops grown on a commercial basis in the United States have been modified to tolerate certain proprietary herbicides or weedkillers. Crops such as corn, soy and canola have been genetically engineered to withstand otherwise lethal doses of chemical pesticides. Farmers, therefore, can douse their fields with herbicides without having to worry about killing their crops.

Herbicide-tolerant crops contradict the claim that genetic engineering will help the environment. Instead of moving farmers away from their dependence on chemical pesticides, the crops actually encourage pesticide use – a threat to our drinking water, our food and to wildlife. A report by Benbrook Consulting in July 1999, which reviewed more that 8,200 university-run field tests on herbicide-resistant crops, found that farmers planting Roundup Ready soybeans used two to five times more herbicides than conventional soybean farmers.

Superweeds

It has been shown that herbicide-resistance genes can spread to related plants in pollen carried by bees or the wind. Researchers have found evidence for this in the case of canola and sugar beet in Europe. In Canada, volunteer canola resistant to three different herbicides resulted from such uncontrollable cross-breeding between plants resistant to one herbicide each.

These superweeds can be difficult and expensive for farmers to eradicate. They could displace existing species of plants, destroying local ecosystems and threatening biodiversity. Research in the UK has shown that pollen from GE crops can contaminate fields up to 4 km away, creating serious liability problems for farmers growing crops for the expanding markets in non-GE and organic foods. Herbicide-resistant crops, themselves, may also become weeds if they grow in places where humans don't want them.

Pesticide Plants

Second in acreage only to herbicide-resistant crops,

pesticide plants are engineered to produce toxin in all their tissues, including the edible grain. Pesticide plants are produced by means of a "gene gun," which is used to "shoot" a toxin-producing gene taken from a soil bacterium – Bacillus thuringiensis (Bt) – directly into the tissue of corn, canola, potatoes and cotton to make the plants poisonous to insects. About 25% of the U.S. corn crop is now planted in Bt varieties. Proponents of genetic engineering argue that Bt crops will reduce the need for insecticides and therefore spare the environment. In fact, the transformation of plants into pesticides is a terribly misguided development with ominous implications for the health of the ecosystem.

Effects on non-target organisms:

The toxin gene found naturally in Bt bacteria produces an inactive "protoxin" that is activated by the gastric juices of certain insects; the activated toxin then destroys their digestive tracts and kills the insects. In contrast, genetically engineered plants produce an active toxin that does not require activation. In 1999, scientists at Cornell University revealed that pollen from genetically engineered Bt corn can kill Monarch butterflies. The findings of this lab study have since been confirmed in an ongoing field study at Iowa State University.

New research also shows that the Bt toxin can leach through plant roots into the soil where it binds to soil particles and remains active for up to 250 days, possibly harming soil micro-organisms and disrupting the soil ecology. Evidence shows that Bt crops may also affect beneficial predator insects such as lacewings and ladybirds when they eat insects that have been feeding on genetically engineered plants.

Threat to Organic Agriculture

Bt crops also pose a threat to organic agriculture. Organic

farmers have long used the Bt toxin in a natural spray as a component of an integrated pest management scheme. The spray targets specific pests and is non-toxic to mammals. However, plants that continually produce the Bt insecticide will exert strong selective pressure on insects to develop resistance to the insecticide. As insects evolve resistance to Bt toxins, organic farmers will be stripped of one of their most valuable tools.

Many scientists believe that efforts to slow the development of resistance in pests are ineffectual. The main strategy involves planting "refuges" of non-Bt crops interspersed among the genentically engineered fields, with the goal of diluting resistant insect populations with susceptible individuals.

3

Regulation of Genetically Modified Plants

During the past decade the biological sciences have experienced a period of unprecedented progress, and nowhere is the excitement of this new era more apparent than in the field of plant physiology. Innovations such as the recombinant DNA techniques are providing new tools for understanding how environmental signals and hormones regulate gene expression and development.

Special attention has been given to the molecular basis of cell cycle regulation of plant growth and development and plant biotechnology. This fast, near dazzling pace of discoveries shows how enormously the subject has changed to embrace the techniques of molecular biology and is reflected in our observation that the index of the current book covers aspects of plant growth regulation which were neither recognisable nor measurable 10 years ago.

Some consumers incorrectly believe that in the United States, the federal government scrutinizes and directs every step of the development of these new plant varieties. Others believe-also incorrectly-that there is no federal oversight of this process. In fact, the U.S. government regulates the field testing-the growing of crops-of plant varieties developed through biotechnology. In addition, the federal government has provided guidelines to plant researchers to assist them in the development of these plants and to address safety questions. Researchers must prepare comprehensive data to support the safety and wholesomeness of new crop varieties developed through biotechnology. This process often requires many years of laboratory and field testing before a product is brought to market.

Food and Drug Administration

The U.S. Food and Drug Administration (FDA) is the primary agency responsible for ensuring the safety of food and feed products. In May 1992, FDA published in the Federal Register a policy statement regarding its role in the regulation of new plant varieties. This document stated that the characteristics of a food, not the method used to produce the food, form the basis of FDA's role in ensuring the safety of foods from new plant varieties.

Thus, FDA's role in the regulation of food is driven by the food product, not the process by which it is produced. This position is not new. In fact, it is based on years of agency experience. Consistent with FDA's "product, not process" position, foods developed through plant biotechnology are judged to determine equivalence to foods developed through traditional plant breeding. "Equivalent" in this context means that there is no meaningful change in nutritional value or composition of the food and that the new variety is as safe as the existing varieties already in commerce.

Included in FDA's May 1992 Federal Register statement, are guidelines to assist plant varieties to ensure this equivalence. These guidelines, or "decision trees," pose safety questions and make recommendations as to when a manufacturer may need to consult with FDA on questions ranging from natural plant toxicants to nutrient content to possible allergenicity. For example, if a company develops a virus-protected squash, FDA recommends that the developer verify that the composition of the squash-including its nutrient content-is substantially unchanged, and that no known allergens are introduced into the squash.

FDA's role in the development of crops through plant biotechnology is one of consultant in all stages of crop development through plant biotechnology. In most instances, FDA approval of new plant varieties-as with traditionally bred plant varieties-is not required because the plants are equivalent to those already in commerce. Based on the questions and issues posed by the FDA decision trees, the plant developer seeks agency consultation as indicated. FDA approval is required if the plant developer does not satisfy the questions posed in the decision trees. For example, if the new plant contains a known allergen that it previously did not contain, FDA requires product approval before it reaches consumers. Or if the new plant contains a food additive not "generally recognized as safe" (GRAS), then FDA approval of the plant is required.

The U.S. Department of Agriculture (USDA) regulates agricultural products and research-including the development of new plant varieties-primarily through its Animal and Plant Health Inspection Service (APHIS). To ensure that new plant varieties pose no threat to production agriculture or to the environment during cultivation, APHIS regulates the research development by requiring permits for the field testing, shipping and delivery of any seed or plants modified through biotechnology.

For example, if a company wishes to develop a plant that is tolerant to a specific herbicide, APHIS must review the research design and results and grant the right to grow the plant in a test plot. APHIS considers possible concerns posed by a new plant variety. Outcrossing, or the unintentional breeding of a domestic crop with a related species, is one such consideration. If there is a high potential for a new plant variety to outcross with a weedy relative and transfer of a new trait to that relative could pose a risk, APHIS may not allow a test plot and further development of such a plant. To guard against environmental risks, most plant developers avoid adding traits that could increase the competitiveness or other undesirable properties of weedy relatives.

As APHIS and the developers of new plant varieties gain more experience with certain crops, they have expedited the regulatory review process by implementing a process called notification. Notification status for field-testing is granted for crops that already have been studied and tested extensively and do not demonstrate environmental risks. In most cases, developers of new plant varieties of these crops simply must notify APHIS of their intent to conduct field testing, provided that the testing plan meets performance standards specified for that crop. A lengthy review of additional research data is not necessary, as long as the changes in the plant fall within descriptions set forth and approved by APHIS.

APHIS oversight of a new plant variety remains in force until the developer requests and is granted a determination by the agency that its oversight no longer is needed. Such determinations of nonregulated status have been granted for a half-dozen new plant varieties, including delayed-ripening tomatoes and virus-protected squash.

Environmental Protection Agency

The U.S. Environmental Protection Agency (EPA) regulates

any pesticide that may be present in food and sets tolerance levels to provide a high margin of safety for consumers. Regarding new plant varieties, EPA oversees the development of plants that are able to protect themselves against insects or disease. For example, the B.t. protein produced in corn plants (for protection from the European corn borer) must be approved by EPA. EPA also regulates the use of all herbicides. As part of this responsibility, EPA regulates the use of a herbicide on a new plant variety that is tolerant to that specific herbicide.

Most products of plant biotechnology have regulatory oversight by at least two, and often three, federal agencies. For example, a potato developed to contain a higher-solids content will be regulated by USDA for field testing safety. In addition, the potato developer must complete a consultation process with FDA. However, if a potato is developed with an insect protection, EPA will be involved in the regulatory process, along with USDA and FDA.

Regulation and Labeling

The United States has several agencies that are responsible for regulation of food. The U.S. Food and Drug Administration is the primary agency regulating food safety. The U.S. Department of Agriculture, which oversees agricultural research and products, and the Environmental Protection Agency, which regulates substances released in the environment, also may play a role in regulating the development of food biotechnology products.

Foods developed through biotechnology will be subject to the same regulatory requirements the FDA uses to guard the safety of all foods in the marketplace. In May 1992, the agency issued specific guidelines for plant-based genetically-modified foods. These guidelines state:

— Genetically-modified food products will be regulated in the same way as foods produced by other means.

— These products will be judged on their individual safety, allergenicity, toxicity, etc., rather than the methods or techniques used to produce them.

— The FDA will evaluate the safety of new ingredients added to foods through biotechnology the same way it now evaluates a new food additive such as a preservative or food color added to a cake mix or soft drink.

Under guidelines set forth in the federal Food, Drug and Cosmetic Act, the FDA has authority over genetically modified food products, drugs for humans and animals, and medical diagnostic devices. Already, the agency has approved for commercial sale products made through genetic modification, including several drugs, vaccines and medical diagnostic devices, and one food processing enzyme used to make cheese.

Safety assessment. The FDA and U.S. food laws place the burden of proving the safety of all new food products on manufacturers. FDA has issued special guidelines to help companies determine whether products developed through biotechnology need premarket government approval. Before a new biotechnology food product can be introduced, the food company must address a number of questions. They include:

— Has the concentration of naturally occurring toxicants or allergens in the product changed?

— Have the levels of important nutrients changed?

— Could any alterations affect the product's digestibility?

— Have accepted, established scientific procedures been followed in the product's development?

— What are the environmental effects of the product's growth and production?

— Does the genetically-modified plant have a history of safe use in food?

— Has the maker introduced into the product any substances that do not have a history of safe use in food?

If any substances are not generally recognized as safe, they must be approved as food additives before the product can be sold. However, the FDA can ban a product from commercial sale at any time if it determines there is a "reasonable possibility" that a material added to a product makes the product unsafe for public consumption. If such a question arises after a product is on the market, the FDA can remove it from stores until the questions are answered.

Premarket approval. The FDA will not always require a company to receive special approval before introducing a new food product or a genetically-modified product on the market. Premarket review and approval will be necessary only under certain conditions. They include:

— If the product has been genetically modified to contain a substance known to cause allergic reactions.

— If the product contains material from a source not currently in the food supply.

— If introducing or removing a substance causes the product's nutritional value to change.

In such instances, the FDA could approve the product, but require it to carry a special label describing its content and characteristic, and health or safety information relevant to consumption of the food product.

The U.S. Department of Agriculture has several roles in agricultural biotechnology. The USDA has broad statutory power to regulate agricultural research and agricultural products, and its guidelines control everything from genetically-modified plants to animal vaccines. The

USDA's Office of Agricultural Biotechnology (OAB) coordinates all the department's biotechnology activities, but 11 of its various agencies participate in some aspect of its work.

Perhaps the single most important of these agencies is the Animal and Plant Health Inspection Service (APHIS), which manages and enforces all USDA biotechnology-related regulations. APHIS also reviews and approves field research projects involving the genetic modification of plants via a plant pest. Potential projects include:

— Plants with new traits for improved disease resistance.

— Bacteria with a new gene that helps plants absorb nitrogen fertilizer.

— An animal vaccine made with genetically-modified bacteria.

In March 1993, after a 60-day public comment period, APHIS announced an amendment to its rules that excludes six crops with a history of safe genetic modification from the prior approval requirement. The amendment allows genetically-modified corn, cotton, potato, soybean, tobacco, and tomato plants to undergo field tests with 30 days advance notification to APHIS. Field test permits granted for these crops between 1987 and March 1993 accounted for 85 percent of total APHIS test permits. Considering the extensive data obtained from these field tests, APHIS determined that the tests of these crops do not result in any substance persisting in the environment or the introduction and dissemination of a plant pest. The amendment provides continued oversight, yet allows the agency to focus resources on applications for tests without such a proven record.

When an application is filed, the agency reviews the genetic background of the material to be tested and the plans for the test. If APHIS approves, it then:

— Establishes the rules under which the test must be conducted.

— Grants the necessary permits if the genetically-modified material is to be transported interstate.

— Gives final approval for the test.

FIFRA

FIFRA is the law that regulates the registration, manufacture and use of all pesticides. Under this law, EPA must review and approve applications for genetically-modified pesticides or crop plants that contain pesticidal properties before any field tests can be conducted.

TSCA

Regulation of genetically-modified chemical products is equally strenuous under the Toxic Substances Control Act. EPA must review and approve every new chemical product that falls under TSCA jurisdiction before the product may be manufactured for commercial use. This pre-manufacture notification process is required even if the product is to remain in a laboratory, fermenting vat or other type of closed system.

Labeling of Foods

Food biotechnology can be used to increase the quality and quantity of food through development of new plant varieties, food additives and food processing aids. These benefits are developed through the use of techniques that represent a continuum from plant breeding to the use of recombining DNA.

The central purposes of food labeling are to inform and educate consumers to enable them to wisely choose food and improve their health. The issue thus becomes not whether a particular technique was employed in the production of a food but, rather, whether nutritional or compositional issues arise in light of that technology, which through labeling, should be brought to the attention of the consumer.

The FDA's labeling policy requires that any issue related to safety must be labeled. Known allergens must be labeled. The policy also requires that when properties of a food or its preparation have been altered, this information be presented to the consumer. Labeling issues are complex. Not only does the system of state and federal laws and regulation mesh to protect and inform the consumer, but also they protect interstate commerce. As the U.S. economy grows more global, one of the country's primary export engines is the agricultural sector. Labeling issues are, indeed, global in nature and actions taken at the state level can have repercussions in the global marketplace.

For this reason, the concept of substantial equivalence is not only discussed in federal regulations, but is being discussed and implemented in the international arena as well. The concept is that if a food is nutritionally and compositionally equivalent to that of its traditional counterpart and its intended use is the same (within the normal range for that counterpart), then the food is judged to be substantially equivalent. Special labeling is not required for such food. Voluntary labeling that is truthful, not misleading, and that is verifiable could be appropriate. Substantial equivalence also encompasses the following concepts.

- The concept of substantial equivalence does not mean that foods or ingredients are "identical." If a food or food ingredient is compositionally and nutritionally equivalent to the traditional counterpart within historical variation limits, it may, at least scientifically, be considered equivalent.

— If the foodstuff is compositionally equivalent to the counterpart but has "well-defined differences," the differences must be subjected to well known nutritional and safety analysis. If there are no implications to the consumer from this perspective, the food may be deemed equivalent.

— If a whole food is judged to be substantially equivalent the fractions of it also would be substantially equivalent. The minute presence of a new protein which, in and of itself, poses not allergenic or other safety risk and which has no meaningful impact on the overall protein content or amino acid profile, would not, as a matter of sound science, change the "characteristic" of the food or alter its equivalency.

Most companies that are developing improved crops and foods believe that the consumer has the right to truthful, nonmisleading information about the composition, nutrition, safety and quality of the foods they buy. Information generally is available to consumers in the food supply chain.

Seed Laws

State seed laws have not been reviewed for some time. Reviews of seed laws may be necessary to accommodate new developments in seed technology. Issues to consider include the following.

— Protect farmers from mislabeled or illegal products. What happens when a farmer receives seed from a conditioner that is mislabeled as Roundup Ready® herbicide, the crop will be destroyed. How does the state prevent farmer liability exposure to this risk?

— Protect seed conditioners by standardizing record-keeping systems. What happens when a conditioner is asked to perform or is accused of violating seed or patent laws or regulations? How can the conditioner minimize the possibility of liability?

— Improve industry communication and stewardship. How can the state contact and communicate with seed conditioners? How can the state ensure proper identity of all seed according to label and the law?

— Ensure profitable technology access for all farmers. New seed technologies have, to date, remained available to all farmers. However, what happens if a few farmers abuse the technology and obstruct the ability to bring new products that offer even more?

Integrated Pest Management (IPM)

Agricultural biotechnology is being used to provide alternatives and different ways to control pests. Because of these applications biotechnology's role in integrated pest management frequently is considered as IPM moves from being defined to being used in production agriculture. IPM is a practice where pest management is but one component in the overall crop production system. The control of pests is viewed as a component of the overall cropping system rather than an end to itself.

The philosophy of IPM programs is based on the principle of giving growers the widest array of options available to control pests in precisely the right amounts at precisely the appropriate times. The program's ultimate goal is to assure production of abundant, high-quality food and fiber in an environmentally and economically sound manner. IPM programs take the view that inappropriate application of a pest control strategy is as much a failure as is failing to control pests when an obvious control need existed. Both prevent growers from achieving the twin goals to provide cost efficient production of abundant, quality food while protecting the environment.

Sometimes IPM is confused with organic production or is used interchangeably with the concept of sustainable agriculture. The rules of organic agriculture vary from state to state. The USDA is preparing rules for organic production domestically, while the Codex Alimentarius is preparing rules for use in trade internationally. Organic production rules tend to adopt methods of practice that consumers and producers regard as Organic. IPM clearly can be a component of organic production, but not all IPM approaches would be organic production methods.

Biotechnology should be an integral component of IPM approaches because of the benefits the technology provides. These benefits include the enhanced use of biologicals, expanded host ranges, supplying protection to hosts that might not ordinarily be protected in that way, different delivery systems, and more specific pest targeting at earlier stages. Currently, approvals of plant pesticides by the EPA involve the use of best available management plans to ensure protection of the pesticidal trait and protect the environment. State legislatures can serve the public interest by participating in the federal program at the state and regional levels. By encouraging innovative research and applying new technology, both the producers and consumers of the state benefit.

Intellectual Property Rights

This is an evolving policy area. Legislative oversight is needed to ensure that intellectual property rights are protected to guarantee that beneficial new crop varieties are available to American farmers. Many of the benefits that biotechnology brings to agriculture are through improving seeds. By placing new traits into seeds, farmers have access to new technology that helps them produce abundant, safe food in economical and sustainable ways.

Agricultural biotechnology developments are more like the development of new pharmaceuticals than like the agricultural developments of the past. It can take a decade and hundreds of millions of dollars to bring new bioengineered variety to the marketplace. This level of investment and that type of research and development produce benefits and policy issues alike.

Protecting intellectual property rights (IPR) is in the best interest of the entire agricultural industry, from growers and retailers and, ultimately, to the consumer. Without IPR, development investment will decline and agriculture's ability to feed a rapidly growing world will be severely affected. The ability of companies to share in the value created by biotechnology will affect their level of investment. Continued access to the technology requires states to update seed laws to protect access and prevent theft of technology that has been properly licensed to growers. Liability protection also must be given to growers, seed conditioners and others in the seed sale chain to protect their access to the technology and to protect honest brokers from fraud and abuse.

Pros and Cons of Agricultural Engineering

Agricultural biotechnology affects the economy, the environment, and the health and safety of the human population as well as animals and plants. The practical and theoretical difficulties center on identifying the long-term, desirable results from the possible but unidentified undesirable consequences. The proponents of using genetic engineering to enhance agricultural, plant and animal products argue that they are only shortening the time frame in which nature and man have done exactly the same thing through multiple growing seasons.

But now instead of waiting for a desirable spontaneous mutation, man can manipulate. Instead of spending years breeding countless generations of cows to achieve the best milk producer or generations of bulls to get the best breeders, man can manipulate. And genetic manipulation of plants can produce changes that might seldom occur or never be noticed in nature in the useful and foreseeable future. Proponents also point out that:

— Genetic engineering carries the potential for increasing food supplies and food nutritional quality in a world where large numbers of people still starve to death.

— By modifying plants to increase the presence of natural pesticides or the resistance to herbicides, the use of toxic and polluting pesticides and herbicides can be decreased.

— Nitrogen fixing enzymes can decrease the need for water polluting fertilizers and perhaps assist in modifying undesirable climate change.

— Plant resistance to different kinds of stress, such as drought, can be altered, thereby improving crop production and decreasing the effects of uncontrollable variables such as weather.

— By exploiting the attribute of some bacteria to digest pollutants, bioremediation can be accelerated to clean up contaminated and unusable sites.

— The medicinal properties of plants such taxol (a cancer fighting drug that originated in the Pacific yew) can be replicated and enhanced in the laboratory.

The arguments against the use of genetic engineering focus on the unknown and the possible risks.

— Might the augmentation of natural pesticides have the potential for increasing the natural evolution of insect pests, which have shown themselves to be very adaptable or will the use of this technology extend their usefulness?

— Might a change or increase in natural plant pesticides have an adverse effect on other species for which the plants provide a benefit in terms of diet or habitat?

— Might an insect resistant plant be so successful that other varieties will die out and essentially all the farmers' products rely on one plant variety? Will genetic engineering decrease biodiversity and unknown plants like the Pacific yew become extinct because they have no apparent economic value?

— Might there be unknown symbiotic relationships that could be eliminated or changed by the introduction or rearrangement of genes? It may be that x is necessary to get y and if x is changed, then y, a desirable attribute, will suffer?

— Might a new attribute be introduced inadvertently and be toxic to some people with allergies?

— If a crop such as rice is grown that provides a human protein to treat cystic fibrosis or to make drugs, could the changes in this crop have any detrimental effects if they appear in the normal food chain?

— Can any of these genetic changes escape and create an adverse effect if they are in mass production and crossbred with other agricultural or uncultivated products?

Regulation of Genetically Modified Pest-producted Plants

Pest control is a continuous process: as pest-protected plants are bred or new chemical pesticides are developed, pests evolve to overcome these control methods. Early methods to control pests include the use of sulfur fumigation in 1000 BC, ants for biocontrol in 324 BC, and crop rotation, controlled irrigation, and manure application during the Roman Empire. Arsenic was used in the 1600s, Bacillus thuringiensis was developed as a microbial insecticide as early as 1938, and the use of synthetic pesticides became the predominant means of pest control in the 1940s.

Since the 1960s, there has been wide implementation of integrated pest-management (IPM) approaches, designed to use a variety of natural controls and cultural methods to suppress pest populations. IPM is an approach which manages pests by biologically integrated alternatives for pest control (US Congress 1947, as amended in the 1972 Federal Environmental Pesticide Control Act, section 136r(a)) and is "a sustainable approach to managing pests by combining biological, cultural, physical, and chemical tools in a way that minimizes economic, health, and environmental risks". Pesticides are used only as necessary and when other control methods have failed.

Genetic Modification Methods

To develop pest-resistant or tolerant cultivars, plant breeders have taken advantage of natural genetic variation or induced mutations. The methods that plant breeders use depend on the type of cultivar they want to improve (for example, an inbred line, a hybrid, or a population) and the reproductive biology of the plant (for example, self-pollinated or cross-pollinated).

An inbred line (or purebred) is phenotypically uniform, and the progeny are identical with the parent. Many self-pollinated crops are released as inbred lines (for example, soybeans, Glycine max, and barley, Hordeum vulgare). A hybrid is the cross between two or more inbred lines; it can also be phenotypically uniform but not genetically identical with the parents. Many cross-pollinated crops are released as hybrids (for example, corn or maize, Zea mays). A plant population results from crossing a number of lines and is genetically and phenotypically diverse, although for key traits, a population can be phenotypically uniform (for example, every plant resistant to a pest).

All genetic modification methods for crop improvement consist of introducing variation, selecting useful variants, and field-testing the selected lines, hybrids, or populations to determine their merit. In the past, almost all commonly used plant breeding techniques began with artificial crosses, in which pollen from one plant is transferred to a reproductive organ of another, sexually compatible plant. Crossing allows for the combining of desirable traits, such as pest resistance and increased yield, from two or more plant cultivars. The objective is to combine these traits in a new cultivar that is superior to its parents.

To overcome some of the barriers to sexual hybridization between cultivated and wild relatives, rescue of pollinated embryos has been used: when a cross yields a viable embryo but the surrounding seed endosperm is not viable, the embryo is taken from the nonviable seed environment and "rescued" by being grown in tissue culture.

Other techniques to introduce variation in cultivars include cell fusion, somaclonal variation, chemical or x-ray mutagenesis, and genetic engineering. Cell fusion is used to produce novel combinations of genomic material from nuclei and organelles when plants are not sexually compatible; it can be performed only on plants that can be cultured with protoplast technologies. With protoplast technologies, cells are disconnected from tissues, their walls are removed, and their membranes are prepared for fusion.

Somaclonal variation is variation that occurs during the tissue-culture process, and its phenotypic outcomes are often similar to other forms of mutagenesis. Genetic engineering is the transfer of a or a few genes into a cultivar with the use of Agrobacterium tumefaciens, microprojectile bombardment, electroporation, or microinjection.

One of the main differences among the techniques used for introducing variation is in the amount of DNA involved. In progeny resulting from a cross between two cultivars, half the genome comes from each cultivar. Each half (haploid) genome contains a significant amount of DNA (table 1). The amount transferred with conventional breeding in the case of Arabidopsis could be 70 megabases (Mb) (half the progeny's haploid genome comes from each parent). For bread wheat, the amount of DNA could be almost 8000 Mb. In contrast, transgenic methods involve the addition of only a few genes and flanking regulatory sequences (totaling about 1-20 kilobases).

Another important difference among the techniques can be the source of the transferred DNA. Sexual hybridization involves genes from sexually compatible species, which tend to be rather similar. Mutagenesis and the somaclonal variation process do not add genes, but rather modify existing genes. Cell fusion can add genes from evolutionarily divergent plant species (such as, plants from different genera), but normally fused cells are from somewhat related plants (for example, the technique has not been conducted by fusing cells from plants and microorganisms).

In genetic engineering or transgenic methods, genes from any organism in the biosphere can be used as long as the regulatory sequences are functional in the host plant. For example with genetic engineering researchers have added genes to potatoes from bacteria, viruses, chickens, and moths. The foreign gene can also be modified by molecular techniques before introduction into the plant (for example, by incorporating DNA base pair substitutions). However, a key question is whether the fact that genes can be obtained from broader sources for plant biotechnology inherently impacts the safety of the resulting genetically engineered organism. Foreign genes engineered into plants may or may not be homologous to genes already present in the plant or the food supply.

Table 1. Genome Size of Common Plants

Common Name	Scientific Name	Nuclear DNA Content: Picograms (diploid nucleus)[a]	Nuclear DNA Content: ~Millions of base pairs (haploid nucleus)[b]
Arabidopsis	Arabidopsis thaliana	0.3	145
Barley	Hordeum vulgare	10.1	4873
Brussels sprout	Brassica oleracea ssp. gemmifera	1.3	628
Corn	Zea mays	4.75-5.63	2292-2716
Cotton	Gossypium hirsutum	4.39	2118
Oats	Avena sativa	23.45	11315
Papaya	Carica papaya	0.77	372
Peanut	Arachis hypogeae	5.83	2813
Rice	Oryza sativa	0.87-0.96	419-463
Soybean	Glycine max	2.31	1115
Tobacco	Nicotiana tabacum	8.75-9.63	4221-4646
Tomato	Lycopersicum esculentum	1.88-2.07	907-1000
Bread Wheat	Triticum aestivum	33.09	15966
Wild Wheat	Triticum monococcum	11.92	5751

[a] 1 picogram = 965 million base pairs, haploid nucleus

[b] DNA content of unreplicated haploid chromosome complement

Breeding Methods

Selection for desirable traits and hybridization has been used since the advent of human agriculture, but the logic underlying the inheritance of traits was not discovered until the middle 1800s. In the 1860s, Gregor Mendel demonstrated the process of heredity by hybridizing different varieties of pea (Pisum sativum) and examining traits such as flower and seed color, seed and pod shape, flower position, and plant height in subsequent generations. His revolutionary experiments paved the way for modern agriculture by showing that through controlled pollination crosses, characteristics are inherited in a logical and predictable manner. In 1905, Roland Biffen, of England, built on Mendel's experiments by illustrating that the ability of wheat (Triticum aestivum) to resist a rust fungus could be passed to later generations.

Since then, many plants have been bred to include desirable traits, such as pest resistance. Blight resistance traits from a Mexican potato species (Solanum demissum) have been introduced into over 50% of all potato cultivars. Blight-resistant corn (Zea mays), rust-resistant wheat (T. aestivum), and aphid-resistant alfalfa (Medicago sativa) are other notable examples of conventional plant breeding.

Major gains in crop yields have been attributed partially to advances in classical plant breeding and plants developed for pest resistance. Corn yields have increased from 5 metric tons per hectare in 1967 to 8 metric tons per hectare in 1997, cereal harvests have been increasing at an average rate of 1.3% per year, world food production has doubled since 1960, and agricultural productivity from land and water use has tripled.

Conventional breeding will likely continue to play an essential role in the improvement of agricultural crops. However, many believe that traditional breeding methods will not be sufficient to meet increasing demands in developing countries for staple crops, such as wheat (T. aestivum), rice (Oryza sativa) and corn (Zea mays). Classical methods are time-consuming (that is they take approximately 10 years to develop a variety) and labor-intensive (only one line of thousands becomes a useful variety).

In addition, beneficial traits can be linked to or lead to undesirable traits, such as disease susceptibility. For example, when male-sterile corn was extensively grown in the 1960-1970s to promote hybrid-corn production, a new race of southern corn leaf blight fungus (Helminthosporium maydis) evolved which successfully attacked this type of corn and significantly decreased US corn yields. Some have proposed transgenic methods to augment the advances in conventional breeding.

Recombinant DNA and Transgenic Pest-Protected Plants

In the past two decades, scientists have focused on expanding genetic modification methods to include the use of recombinant DNA (rDNA) techniques. New varieties generally can be produced faster by rDNA than by conventional breeding methods. rDNA methods allow the introduction of genes from distantly related species or even from different biologic kingdoms.

In addition, detailed knowledge of the trait being introduced (such as a DNA sequence or cellular function) can lead to less variability in the offspring and eliminate some of the uncertainty about linked traits. The site of insertion of a gene can affect its expression, but generally plants with the appropriate level of expression can be selected if a number of transgenic plants are produced. After a trait is introduced by transgenic methods, the resulting plant can be sexually hybridized with useful varieties developed by conventional breeding.

Recombinant DNA methods emerged in the early 1970s after the discovery of restriction enzymes, DNA-sequencing methods, and plasmid and viral vectors for engineering organisms. The methods have been used ever since to manipulate DNA fragments that contain genes of interest. With the advent of this technology, concerns about the safety of experiments that use rDNA methods developed. The National Academy of Sciences (NAS) in 1974 convened a committee to assess the safety concerns associated with rDNA research. The committee recommended that rDNA experiments be postponed until further evaluation of the risks.

Soon after, the International Conference on Recombinant DNA Molecules, better known as the Asilomar Conference, was held. An outline of guiding principles and restrictions for rDNA research was generated at this conference. In 1976, the principles were reviewed by the National Institutes of Health (NIH), which implemented official guidelines to be administered by the NIH Recombinant DNA Advisory Committee (RAC). The guidelines focused on laboratory containment of rDNA microorganisms. As more experiments were conducted and more data on the risks were generated, less restrictive guidelines were put into place. In recent years, the guidelines have been expanded to include other rDNA applications, such as gene therapy, and have been adopted not only by institutions receiving federal funding, but also by industry and state institutions.

About a decade after the emergence of rDNA technology, genes from the bacterium Agrobacterium tumefaciens were used to carry foreign genes into plants. Agrobacterium inserts portions of its tumor-inducing (Ti) plasmid-encoded genes into plant chromosomes as part of its natural, parasitic life cycle. When researchers add foreign DNA (such as a gene for pest-protection) in between Ti plasmid-encoded insertion sequences, the foreign DNA sequences are also inserted into the plant's chromosome.

The first transgenic plants were developed with Agrobacterium-transformation methods. Since then, other methods for plant transformation, such as electroporation and particle-gun transformation, have been developed; these methods allow transformation of plants that are not natural hosts for Agrobacterium.

Regulatory Framework for Transgenic Plants

Concurrently with developments in the technical aspects of genetically engineering crops by using rDNA methods, regulatory concerns about the release of genetically engineered organisms into the environment emerged. Under that regulation, coverage extends to organisms or substances that meet the definition of a plant pest or that USDA chooses to designate as a plant pest.

Table 2. Summary of the Current Regulation of Biotechnology Products, as currently described on the USDA website for Regulatory Oversight in Biotechnology.

Agency	Jurisdiction	Laws
US Department of Agriculture	Plant pests, plants, veterinary biologics	Federal Plant Pest Act
Food and Drug Administration	Food, feed, food additives, veterinary drugs, human drugs, medical devices	Federal Food, Drug, and Cosmetic Act
EPA	Microbial and plant-pesticides, new uses of existing pesticides, novel microorganisms	Federal Insecticide, Fungicide, and Rodenticide Act; Federal Food, Drug, and Cosmetic Act; Toxic Substances Control Act

Source: USDA 1999e.

First Field Trials

In November and December 1987, USDA issued permits for three engineered herbicide-tolerant varieties of tomato (two from DuPont and one from Calgene) and two herbicide-tolerant varieties of tobacco (from Calgene). These plants were tolerant of the herbicides glyphosate, bromoxynil, or sulfonylurea. Tolerance was based on the overexpression of the herbicide target in the tolerant plant, the expression of resistant forms of the target enzymes, or the expression of enzymes that could degrade the herbicide.

Transgenic pest-protected plants were developed in parallel to herbicide-tolerant plants. The first transgenic pest-protected plant was engineered to contain a coat-protein gene from the tobacco mosaic virus (TMV); the gene confers resistance to TMV itself, and to viruses similar to TMV. A transgenic TMV-resistant tomato line developed by Monsanto was approved for field trials on March 23, 1988.

Today, a large portion of US corn and cotton acreage is planted with transgenic pest-protected plants. Those transgenic pest-protected plants contain genes from the bacterium Bacillus thuringiensis (Bt). Bt produces several proteins during sporulation including endotoxins. Upon ingestion by an insect, the protoxin form of endotoxin undergoes cleavage in the insect gut to a truncated active form, which kills insects by binding to receptors in the insect gut and forming pores.

The pores cause the gut contents to leak into the blood, and this eventually leads to insect death. About 60 proteins from more than 50 subspecies of Bt have been identified in the last 20 years. Which insects Bt toxins affect depends on the class of Bt protein; they include moths and butterflies (lepidopterans), flies and mosquitoes (dipterans), and beetles (coleopterans). Mixtures of Bt have been used to spray crops for over 50 years. However, the Bt toxin generally loses its effectiveness in the environment within a few days. Sometimes spraying needs to be done frequently.

In transgenic crops, Bt toxin is continuously produced and is protected from the elements. It therefore retains its ability to kill pests during the entire growing season. Moreover, the toxin is generally expressed in every part of the plant, including internal tissues that are difficult to protect with topically applied pesticides. This internal production provides protection against pests that are internal feeders such as the pink bollworm in cotton and the European corn borer in corn. On the other hand, the constant presence of Bt toxin in transgenic pest-protected plants during the growing season has led to concerns about its persistence in the environment and increased probability of pest evolution to overcome the protection mechanism.

The first permit to field-test transgenic crops that contained Bt genes was issued to Monsanto in 1988 for tomato. Initial attempts to make crops that would resist pests in the field were not successful because of problems with the expression of the bacterial genes. In 1990, the first successful Bt crop, cotton (Gossypium hirsutum), was produced by overcoming translational and transcriptional barriers to bacterial-gene expression in plants. Transgenic methods for introducing Bt are often followed by conventional breeding with varieties that express other useful agronomic traits.

Before transgenic crops were commercialized (from 1987 to 1994), the USDA approved field trials of nine nematode-resistant transgenic pest-protected plant varieties, 45 fungus-resistant varieties, 17 bacteria-resistant varieties, 322 insect-resistant varieties, and 194 virus-resistant varieties.

Commericialisation of Transgenic Pest-producted Plants

Under the USDA, FDA, and EPA statutes, the first transgenic crop varieties were approved for commercial planting in the early 1990s. In 1992, the first transgenic crop variety achieved nonregulated status from USDA. This variety was a tomato line for altered fruit ripening developed by Calgene (Flavr Savr). In December 1994, the first transgenic pest-protected plant achieved nonregulated status from USDA: a virus-resistant squash variety developed by Upjohn/Asgrow Seed Company that contained watermelon mosaic virus-2 coat protein and zucchini yellow mosaic virus coat protein. The USDA assessments for this crop address such concerns as the likelihood of creating new plant viruses via recombination of the introduced coat-protein gene with naturally occurring viruses, the potential of the two new virus-resistance genes to cause squash to become a weed, and the movement of the genes to wild squash relatives.

Varieties employing the Bt resistance mechanism were the next pest-protected plants to achieve nonregulated status from USDA and to have their gene products reviewed as plant-pesticides by EPA. A Bt potato line resistant to beetles was developed by Monsanto and was cleared for commercial release by USDA in March 1995, subject to EPA and FDA review. The Cry3A delta-endotoxin from Bt was reviewed by EPA in early 1995.

An exemption under FFDCA from the requirement of a tolerance for this Bt toxin and the genetic material necessary for its production eliminated the need to establish a maximal permissible level for residues of this Bt toxin in potatoes.

Over 40 transgenic crop varieties have been cleared through the federal review processes for commercial use in the United States. Of them, 17 contain transgenes for pest-protection. Of the 17, 14 containing Bt genes have been developed and cleared by USDA for commercial release (table 3). Although the EPA 1994 rule is not yet final, the plant-pesticides in these crops have been reviewed and their gene products registered as plant-pesticides by EPA. Five virus-resistant transgenic pest-protected plant varieties have achieved nonregulated status from USDA (table 3).

Transgenic crops were first planted commercially in the 1995 growing season. Since then, their use has been rapidly increasing. In 1997, 20.3 million acres of transgenic crops were planted in the United States; in 1998, 50.2 million acres were planted; and in 1999, 70 million were planted. A total of 98 million acres were planted worldwide in 1999.

Transgenic pest-protected crop varieties that contain Bt toxin transgenes make up a large percentage of the commercial transgenic crops. In 1998 in the United States, about 25% of total cotton acreage and 21% of total corn acreage were planted with transgenic crops that contain Bt genes. In 1999, approximately 30 million acres of insect protected crops were planted in the US. So far, many of these transgenic pest-protected crops seem to be effective in controlling pests; a reduced need for chemical pesticides and increased yields have been reported by many, but not all, growers.

Table 3. Crops Deregulated by the USDA with Transgenic Pesticidal Traits

Crop	Resistance[a]	Company	Date of Nonregulated Status
Insect resistance			
Potato	Bt IR	Monsanto	March 1995
Corn	Bt IR	Ciba-Geigy	May 1995
Cotton	Bt IR	Monsanto	June 1995
Corn	Bt IR	Monsanto	August 1995

Corn	Bt IR	Northrup King	January 1996
Corn	Bt IR	Monsanto	March 1996
Potato	Bt IR	Monsanto	May 1996
Corn	Bt IR	Dekalb Genetics	March 1997
Cotton	Bt IR, HR	Calgene	April 1997
Corn	Bt IR, HR	Monsanto	May 1997
Tomato	Bt IR	Monsanto	March 1998
Corn	Bt IR, HR	AgrEvo	May 1998
Vir[illegible] and insect resistance			
Potato	Bt IR, VR	Monsanto	December 1998
Potato	Bt IR, VR	Monsanto	February 1999
Virus resistance only			
Squash	VR	Upjohn/Asgrow	December 1994
Squash	VR	Asgrow	June 1996
Papaya	VR	Cornell University	September 1996

[a]Bt IR = Bt endotoxin-based insect resistance; VR = virus resistance; HR = herbicide resistance

However, estimated benefits might depend on the baseline level of pest infestation during a specific growing season and on the techniques used to make comparisons. The use of transgenic pest-protected crops has been profitable in growing regions subject to severe pressure from specific pests or where alternative means of pest control have been infeasible or expensive. For example, Bt cotton has been accepted by a large percentage of growers in states where pest resistance to synthetic pyrethroids has left them without chemical means of controlling bollworms, but limited in other regions where pest hazards are not so extreme. Adoption of Bt corn has similarly been limited to areas with the highest pest pressure.

In addition to the approved commercial transgenic crop varieties, thousands of transgenic varieties are undergoing field trials. From 1987 through January 2000, the number of permits issued and notifications acknowledged was over 6700; about 3000 were for varieties having pest-resistance genes (table 4).

Given the rapid increase in acres planted with commercial transgenic crops and the likely additional increase in their use, many groups have raised concerns about the ecological and human health risks that might be posed by these crops. Although the risks might not, in principle, differ in type from those associated with other products (for example, conventional pest-protected plants, pesticides), the public has focused its attention on transgenic crops.

Concerns over pesticidal traits include the enhanced evolution of resistant pest strains, the toxicity or allergenicity of the gene products to humans, the hybridization of transgenic pest-protected plants with neighboring wild relatives, and adverse effects on nontarget organisms.

Table 4. Number of Permits Issued for or Notifications of Field Trials in the United States Involving Crops with Pest-Resistance Genes, 1987-1999.

Category of Resistance	Number (% of all field trials)
Insect	1505 (22)
Virus	1013 (15)
Fungal	378 (5.5)
Bacterial	78 (1.2)
Nematode	7 (0.1)

Pest Resistance to Engineered Traits

Farmers and gardeners who use microbial Bt sprays are concerned that the widespread commercial planting of transgenic pest-protected plants with Bt genes will lead to rapid development of insect resistance to Bt, which will in turn make their microbial sprays ineffective. Instances of pest adaptation to conventional Bt products have been documented.

Scientists who conduct research on pest resistance to plant-protection mechanisms published resistance management strategies for Bt corn, cotton, and potato, and the EPA published findings of a specially convened scientific advisory panel on Bt resistance management. Under the registration process for plant pesticides, EPA requires a particular amount of non-Bt cotton or corn to be planted next to Bt cotton or corn to serve as a refuge for insects carrying Bt susceptible genes, and they also encourage the development of resistance management strategies for other transgenic Bt crops.

However, the percentage of acreage that is needed to provide a sufficient refuge to avoid the rapid development of pest resistance and the proper location of the refuge are debated by industry, entomologists, and environmental groups. Recently, the EPA placed new restrictions on growing transgenic Bt corn which include a requirement that farmers plant 20 percent to 50 percent of their corn acreage with conventionally bred corn.

Gene Flow and Cross Pollination With Weedy Relatives

Other safety issues which have received attention are those involving ecological risks such as the effects of gene flow. Studies have been conducted to assess the potential for gene flow among and within related species. The ability of transgenic plants to cross-pollinate with their wild relatives and form offspring with enhanced weediness has been investigated when herbicide-tolerant rapeseed plants were back-crossed with a wild relative.

The hybrid progeny plants produced an equivalent amount of seed as the wild genotypes and were also herbicide-resistant. That study indicated that back-crossed generations of hybrids between transgenic and nontransgenic crops can have the same potential to flourish as other plants. In a more controversial study, wild type Arabidopsis thaliana plants were found to be fertilized by pollen from transgenic plants more often than by pollen from nontransgenic plants.

In addition to those experiments, the use of models has been explored to assess the invasiveness of engineered organisms, although indications are that these models will require several years worth of data to be validated.

Nontarget Species

Although some transgenic pest-protected plants have the potential to reduce pesticide use and thus to prevent substantial environmental damage, there is concern that gene products from the plants could harm beneficial insects or birds (nontarget species) that are in direct contact with the plants or that feed on insects that are. Hilbeck et al. found that when chrysopid larvae were reared on prey that were fed Bt-producing corn, they had 62% mortality. When they were reared on prey that were fed non-Bt corn, mortality was only 37%.

Another experiment indicated that Bt toxins can bind to humic acids from soil, be protected from biodegradation, and persist in the soil. It is not known whether nontarget organisms would be affected by bound toxin molecules in field situations. Other studies indicate that Bt toxins generally degrade quickly in the soil.

Many believe that transgenic crops present substantial human health and ecological risks, and that these are not properly assessed by the regulatory framework. But many others believe that the risks are minimal, that the benefits outweigh the risks, and that the current regulatory scheme is perhaps onerous.

Cited benefits include a reported 250,000-gallon reduction in chemical pesticide use in 1996 and a 30-50% reduction in the number of insecticide applications over the period of 1996-1998 due to the growing of commercial transgenic Bt cotton. The reduction might prevent much environmental damage. In addition, Bt toxins have specific insect targets, whereas traditional broad-spectrum chemical insecticides often kill insects more indiscriminately. This may lead to outbreaks of secondary pests requiring the use of more insecticides. However, many believe that transgenic pest-protected plants should not only be compared to the use of chemicals, but also to alternative methods such as biological control.

4

Organic Agriculture

Organic agriculture is an agricultural production method in which yields are at least adequate economically but are also sustainable, respecting the natural resources at or near the production site (for example, water, air, soil fertility, disease resistance). Organic farmers aim for optimal production under the conditions set by organic agriculture organisations. Environment-friendly principles and methods have replaced chemical fertilisers, pharmaceuticals, pesticides and growth regulators. Organic farmers stimulate production through fine-tuned management of the soil. Effective soil management is regarded as the linchpin on each organic farm.

The organic farmer's respect for nature is reflected in his/her actions, which are based on the self-regulating capacity of the various living organisms. Initially, the sector was strongly focused on optimising crop growth, opening trade channels, increasing sales and boosting public image. Not until recently have efforts been made to optimise the basic seed stock. These efforts have primarily been directed at propagation under organic conditions; improving varieties has long been ignored. Now, with the rapid rise of biotechnology, there is a greater urgency to focus specifically on seed stock and breeding. Below, a number of developments in organic agriculture are described against this backdrop.

Most organic farms in the Netherlands find their origin in conventional farms that had given up livestock production and later converted to organic farming. However, organic farming is increasingly aiming at a closed nutrient cycle which can be achieved by combining crop production and livestock. Thus arable farms are now seeking ways to re-introduce mixed systems by letting beef cattle range outdoors in summer and stabling them in winter. Other arable farms are forming collaborations with more or less distant livestock farms: a kind of mixed system at one remove.

Such farms are now calling for new plant breeding products. Mixed farms need, for example, animal feeds that fit the types of animal that are most suited to their farming system, together withlong-strawed cereal varieties that yield sufficient straw to mix with slurry.

Enhancing Cultivation Systems

Enhancing soil condition (soil life, humus content and structure) as a basis for sound crop growth is central to every organic system. It means that a sound fertilisation regime is coupled to a sound cropping programme. The choice of crop is determined by market demand and complementarity of crop characteristics. Thus a one- sided burdening of the soil is prevented.

In organic farming we find greater diversity in crop management systems because organic systems are more dependent on changing natural circumstances than are conventional systems. Organic farmers should be able to choose from a wide range of crop characteristics. Plant breeders should play a major role in all this.

Production Chain

An organic farmer needs shrewd insight and skill to be able to fine-tune the complexity of individual management factors. He must adopt the attitude of the researcher. Conversely, for research to be effective it should be carried out at farm management level. Regional diversity should also be taken into account which makes a close relationship between organic farmer and plant breeder highly desirable. The diversity in crop management systems is caused not only by organic farming's greater dependency on natural variation but also by diversity in markets.

The food processing industry and the supermarkets are increasingly focusing on organic products. The organic sector produces for a variety of market segments: thus large-scale farms focus on exports or the processing industry, medium-sized farms produce for the specialised health food shops and smaller farms produce for local markets or their own customers (via subscriptions). This contributes to the diversity in crop management systems.

The crop varieties for organic farming should be matched to this diversity in crop management systems and should be sufficiently wide-ranging. An increasing number of consumers express their concern for the quality of food, nature and the environment in their choice for organic produce. Initially consumers focused on food safety: they wanted organic products produced without the use of fertilisers and pesticides. This group is to a large extent the same that is now against genetically modified products. The organic sector's explicit decision to remain gmo-free takes away much of that worry.

The concern for genetically engineered food is acknowledged by the Dutch government in their promise that they will support a gmo-free production chain alongside conventional production. The consumer trusts (that) the (organic) sector will not go back on this. The organic sector cannot therefore allow genetic engineering, not even on scientific grounds. The ever growing number of organic consumers (subscription is becoming very popular) shows that the natural image of organic farming makes the consumer feel that organically produced food is safe.

More than half of the organically farmed land is grassland. In organic farming grassland shows the fastest expansion. This is partly because dairy products were among the first organic products to be introduced in supermarkets. This enabled large-scale marketing and made it easier for farmers to take the step towards conversion. Other sectors of organic farming saw an increase in holdings and farmed area.

Platform Biologica, the interest group for organic farming and the marketing of organic products, gives four reasons for the growth in organic farming:

— an active attitude of supermarket chains in the marketing and promotion of organic produce;

— EU hectare premiums for organic farming;

— a more active role of the government in educating the public about organic farming;

— falling consumer prices following increased sales, which resulted in economies of scale in processing and distribution.

The growth of organic agriculture in the Netherlands compares poorly with that in some other European countries where incentive measures were taken earlier, which allowed their organic sector to develop more quickly. Also, conventional farming in the Netherlands is highly productive and specialised. Many farmers had made major investments and for them conversion was a far greater step than for their European counterparts.

Characteristics of Organic Growing

The organic farmer will base his choice of crops on soil type, soil condition, climate, rotation and market demand. More often than not, his criteria will be similar to those of his conventional counterpart but his priorities are different.

In organic growing, qualities such as rapid early growth, (the crop's contribution to weed reduction on the farm), resistance to pests and diseases, adaptability to lower fertiliser input, role in crop rotation and product quality are important. Obviously yield is important too, but it is not the overriding factor. In his selection of varieties, the grower will use the information provided by seed suppliers, trade journals and the experience of his colleagues. As most varieties have been developed for conventional growers who use chemical fertilisers it is not always easy to say how they will perform under organic conditions. Organic growers have to find out by trying. Comparing notes therefore is essential here.

Nevertheless, the most popular varieties with conventional growers are also used by organic farmers. Older varieties do not find much favour as yields and resistance are poor. However, the fact that organic growers use modern varieties does not necessarily mean that they are the best for organic conditions. Important characteristics in varieties for organic farming are described below.

The organic farmer relies heavily on the natural resources available on the farm and wishes to be independent of external artificial means to the largest possible extent. This requires great adaptability of the crops grown. The corrective actions a farmer can apply are also limited. To get the best possible yields on a given site, growers use cultivars that are adapted to that particular environment and to nutrient levels which fluctuate with the seasons. Cultivars argued that , play an active rather than passive role in all this.

This optimal interaction also concerns adaptability to the dynamics of conditions of growth which changes with the seasons. For instance, in early spring when mineralisation is just starting, nitrogen levels in the soil are low. This requires crops with longer taproots which can bring nutrients up from deeper layers. However many new varieties such as the leek hybrids, require heavy nitrogen applications to deliver the promised yields, which shows how the needs of organic farmers are not met by conventional breeding.

Chemical herbicides are not used in organic farming. Thus a crop's vigour and ability to compete with weeds, particularly in its early growth stages, are vital, as good establishment is critical to the success of the crop. Crop structure, tillering (speed and size) particularly under less favourable conditions (cold spring, low nitrogen levels in soil) and a plant's firmness are all qualities that help farmers reduce weeds. Increasing mechanisation of harvesting such crops as carrot has led to the breeding of varieties with upright tops. This is in sharp contrast with the desire for more ground cover to discourage weeds.

'Many modern wheat varieties from conventional breeding start developing late in spring and give poor early growth (and tillering) under organic conditions. This gives unbalanced and open crops. The growth pattern of modern varieties results in reduced ground cover which reduces the plants' ability to compete with weeds. Their greater sensitivity to weed competition requires a more frequent application of nitrogen than with older varieties. Yields of modern wheat varieties turn out to be relatively unstable over the years. Nor is the use of older varieties the better option as older varieties give poor yields and, under current conditions, are extremely susceptible to pests and diseases.

Pests and Diseases

Organic farmers use prevention rather than control measures to head off pests and diseases. This means that resistance in varieties is an important quality; it should be effective and durable. The ability to resist pests and diseases is based on more than a plant's absolute resistance. Absolute, monogenetic resistance is likely to break down sooner or later, as, for example, in apple production where a new race of the apple scab pathogen has been found recently in a newly introduced resistant variety. Such breakdowns can be delayed by resistance management. In cereal growing for instance, mixed varieties can be used, or varieties can be spread or rotated.

From an organic point of view, each of the varieties used in this way should have a broadly-based resistance. In combination with management measures, such varieties should be adequately resistant to diseases and pests. In commercial plant breeding, optimising and managing varieties in such a way has largely been ignored because it was either too expensive to develop (polygenetic resistance) or did not go far enough (conventional farmers want high resistance levels).

Crop structure is another quality that helps keep down the risk of disease. Plants of a more open structure and less foliage will dry sooner, which reduces the risk of fungal infections. In lettuce this results in more upright growing varieties such as the batavia types. A more open structure however makes it more difficult to keep weeds under control. In this respect a grower has to weigh his options carefully before making a choice. Diseases that are common in conventional farming due to high crop densities and frequent nitrogen applications such as mildew and lodging in cereals are rarely a problem in organic farming.

Organic farmers pay much attention to quality characteristics such as taste, keeping quality, form, structure and colour of a product. Quality is controllable when a farmer carefully selects variety characteristics, soil type and management strategy. Now that supermarket chains are increasingly selling organic products, the pressure on prices is high and strict demands are made on quality. New carrot varieties for instance, that meet the demands on size, smoothness and high yields, lack taste. This might be improved by a change in breeding priorities

Firmness and early maturity are important for a good quality product. Sometimes specific characteristics are important, baking quality in bread wheat, for instance, and malting quality in barley. For baby food and juice processors low nitrate levels in the end product is important. Food processors are increasingly looking for organic products and demand improved variety characteristics and optimised management measures.

Seeds, seedlings and plant material are the basis for cultivation. Good quality is vital here. Producing good propagating material requires great skill. In organic farming an effort is made to work with organically produced stock. Most propagating material however is conventionally grown. Seeds treated with conventional products may be used only if users are unable to obtain non-treated seeds on the market . However organic growers are not allowed to use detergents or germination promoters. In arable farming where seed production does not differ much from crop production, organic growers use their own non-treated seeds as basic stock for self-fertilising crops such as cereals, peas and beans.

There are firms that produce organic propagating material and seed stock of various field crops (e.g. potato) but the organic propagation of hybrids has not yet begun. Skal makes an annual inventory for the benefit of affiliated farmers of the varieties for which organic propagating material is available. In the vegetable sector, organic propagation is a new development. A new foresight study of the problems and possibilities for organic propagation in organic vegetable growing has shown that in this respect conventional seed merchants can make an important contribution.

In the Dutch organic sector there are only two firms specialising in the production of organic seed stock. Guidelines for the organic propagation of seeds vary within Europe. In Switzerland, seed stock can only be traded under the name of Demeter (biodynamic mark) after having run a two-year cycle on an organic farm. The Netherlands follows Council Regulation (EEC) No 2092/91 which allows the use of the EKO mark for selected seeds that under current conditions have been multiplied for one cycle on an organic farm.

Under the farmer's privilege, farmers are allowed to propagate seed stock for their own use. Biodynamic farmers in particular are interested in producing their own seed on the farm as they feel that propagation and selection on their own farm will increase a variety's adaptability to local production circumstances. Adaptability to local conditions requires some genetic variation in a plant but genetic variation in modern varieties is often small. Such varieties are not flexible in adapting to changes in the production site's environment.

The propagation of seed stock on the farm still needs to be translated into a practical strategy. Most farmers cannot meet the requirements set for regular seed quality. Besides, the progeny of selected plants requires lines of certain quality; F1 hybrids, of course, cannot be propagated directly. The propagation of varieties on farm requires expert guidance to carry out proper selections.

Not many farmers therefore think that propagation of seed stock on the farm is a viable option. However, cereal growers are an exception and some growers have years of experience with propagating their own stock. What is important here is that propagation should be under organic conditions. This is increasingly the case as farmers propagate and trade seed stock from one or sometimes several crops on the basis of a contract and with the help of a seed producer (which in the Netherlands is beginning to take off) or in co-operation with growers associations.

Organic Plant Breeding

The fact that conventionally grown varieties are used in organic farming does not necessarily mean that they are the most suitable. With respect to nitrogen uptake efficiency, contribution to weed reduction on the farm, resistance to pests and diseases and quality they leave much to be desired. An organic sector in development needs plant breeding. The growing impact of gene technology in conventional breeding must also be reckoned with. The organic sector's stated intention of wanting to steer clear of genetic modification implies that it does not wish to be dependent on conventionally produced varieties. The sector sets strict requirements for a variety's characteristic (product quality) and breeding techniques (process quality).

Plant breeding programmes that are focused exclusively on organic farming hardly exist on the scale needed, but the need for such programmes is rapidly increasing. In Germany and Switzerland, such programmes have already been set up by breeders from the biodynamic sector. Now that a start has been made with organic propagation in the Netherlands, attempts are being made to set up organic breeding programmes. At the request of organic traders, the Zaadgoed Foundation was set up in the spring of 1998 to stimulate such programmes. The Foundation is a knowledge and coòrdination centre.

Principles for Organic Breeding

Various initiatives in organic plant breeding are aimed at the needs of the organic farmer and at distinguishing organic from conventional breeding in both product characteristics and breeding methods. These initiatives take place in two dimensions:

— plant-environment interaction (ecological dimension)

— the interaction between breeder/seed supplier and farmer (socio-economic dimension)

Like other living organisms, plants are products of evolutionary processes. Evolution is the adaptation of organisms to changes in their environment. The success of organic farmers is related directly to their ability to home in on this adaptive ability. After all, conditions on a working farm cannot be compared to the trial plot set-up where plant breeders evaluate the performance of a new variety or strain. On conventional farms, artificial fertilisers and pesticides are used to copy the "ideal" conditions on trial plots. If a conventional farmer wants to optimise the 'promised' plant characteristics, he must also reduce biotic and abiotic variation on his farm.

On the organic farm, genetic variation should be available to allow crop plants to adapt to the conditions of the individual farm or field. In an organic farmer's eyes, diversity is a valuable resource rather than a hindrance. For example, there may be more beneficial organisms to control those that are harmful. Organic farm management encourages a closed farm system and optimal self-regulating ability, both of which respect natural processes. Organic farmers value the plant-environment interaction

- and the conditions for this interaction - at least as much as the economic performance of a variety or strain.

The objectives of a closed system, self-regulation and diversity at farm level, can be translated to the plant level as:

— natural reproductive ability

— independent adaptive ability

– species authenticity and characteristics

In organic agriculture, it is considered that the ability of plants to reproduce themselves is indicative of vigour, and that vigour is an expression of a plant's health and nutritional value. Thus, organic farmers prefer to use strains which have retained their reproductive ability. Maintaining this characteristic in cultivars is not as easy as it sounds. Many cultivars do not flower and/or disseminate without human intervention. Hybrids, for example, lose their specific combination of characteristics in a next generation.

Examples of cultivated crops which depend on human intervention for flowering and seed production

— Before lettuce can flower, the breeder must cut off the head or pick the leaves so that the flower stem can grow through the head to flower and form seed.

— A cross is cut in white cabbage so that the main stem can grow through the cabbage to flower and form seed.

— In maize and wheat, the kernels are fixed so tightly in the spike or cob that these can no longer disseminate spontaneously. Threshing frees the kernels from the husks

In the examples above, the plants have retained their reproductive potential. With some forms of hybridisation, however, the reproductive ability is lost. In cross-fertilisation, the parent lines of hybrids are artificially inbred. In leeks, for example, the parent lines are no longer able to reproduce themselves and are maintained through tissue culture.

Cytoplasmic male-sterile maternal lines are often used in the production of hybrid seed. These maternal lines are unable to reproduce because they do not produce functioning pollen. This method is the most extreme example of lost reproductive ability and one which also affects the reproductive ability of the cross, in effect damaging the plant's vitality. Moreover, cms hybrids, or cybrids, cannot be used for breeding. From an organic point of view, the production of such varieties is unethical. The plants are used without consideration for the centuries of cultivation that have gone before.

The latest technique is single-use seed. Genes are chemically treated to prevent a next generation of seed from germinating. This technique was developed to protect the interests of investors in the development of genetically modified varieties.

The use of vegetative, or asexual, propagation methods is on the increase. There are crucial differences between asexual propagation and sexual, or generative, propagation. In asexual propagation, all existing characteristics are passed on (more of the same), but there is a risk that diseases are transferred too. Propagation through seed, on the other hand, limits disease transfer so that a new generation can be said to be stronger. A disadvantage of sexual propagation in cross-pollinators is that new generations have different characteristics from the parents, thus requiring a new selection process.

From an organic point of view, there are no major objections to the use of vegetative propagation methods, since this ability actually comes naturally to a number of plants. In fact, in some cases, one is forced to resort to vegetative propagation. However, if vegetative propagation is used solely to secure certain characteristics, we believe that periodic 'refresher' generative phases should be included in the breeding programme to ensure sufficient resistance and vigour.

Participatory plant breeding

Modern breeding, particularly in the developed world, is characterised by a strict division between breeders as the 'creators' of new varieties and farmers as 'users'. At the same time, conventional plant breeding has too little consideration for the diversity in farm conditions, preferring to focus on standard varieties and strains. These limitations are economic, institutional, technical and conceptual in nature.

Economic, because the costs of breeding programmes and the commercial objectives of propagators result in varieties that cover the largest possible geographical range. Institutional, because of the limited possibilities for breeding crops for specific environments. Technical limitations pertain to the limited understanding of genotype-environment interactions. And finally, conceptual problems result from product-oriented technology development that ignores the context in which new products and technologies are to be applied.

These are the limitations of a linear model for the development and transfer of genetic material which starts with the parent lines in breeding programmes and ends in varieties to be used by farmers. The limitations of this model have important ecological consequences. Modern genetic plant matter is developed for highly conditioned production circumstances. Typically, the varieties are tested in circumstances designed to optimise yields. As a result, production conditions in real farm practice have to be standardised, for example by applying chemicals to the land and crops.

In the linear model of seed technology development and transfer, deficiencies in the plant's ability to adapt to 'deviating' production conditions do not become visible until the variety is introduced into farming practice. Current propagating material thus often proves inadequate in organic farming conditions. A variety may be perfect, at least in the short term, for the standardised, maximum yield growing circumstances on conventional farms, but fall short of the mark in organic agriculture.

Developments in organic farming demand other solutions than those offered by the linear breeding model. One commonly heard solution is to develop a basic variety with average characteristics and a large geographical range. Using biotechnological techniques, this basic variety could be adapted for regional production; conditions and characteristics could be accentuated or toned down on demand. Certainly, the result is inter-local diversity, but it is based on 'basic technological developments'. The incorporated adaptations would be relatively simple and monogenetic in nature. A monogenetic approach does not result in fundamentally different propagating material and is not in accordance with the phenotype approach in organic farming.

Rather than genetically improving specific characteristics, organic plant breeding should be a continuous process in which propagating material is adapted to future production conditions. Such a process is the result of a complex interaction between the propagating material and the production environment, a method based on a cyclic rather than a linear model, as demonstrated by Kunz et al.

Varieties are permanently being adapted to local production conditions. The process commences with the selection of parents on the basis of their suitability for the conditions in which the future variety is to be grown, ie. not on the basis of an ideal standard production regime. After the new variety has been formed, selection for farm circumstances/region continues 6 to enable an optimal adaptation of the variety to future growing circumstances. Loss of genetic variation is kept to the necessary minimum for successful growth on the farm concerned.

In the organic approach, the relationship between breeders and farmers changes. In this specific approach, breeders and farmers interact in a cyclic process as they continually strive to develop new propagating material. This is a drastic change from the linear approach, but it is not an entirely new one. In Third World countries, breeding processes based on breeder-farmer interactions are being discussed and developed. This is referred to as participatory plant breeding.

The goal of participatory plant breeding is to develop local varieties adapted specifically to local production conditions. In such a breeding programme farmers receive plant material (not necessarily an officially listed variety) with sufficient genetic diversity to allow selection in, and adaptation to, local conditions. The linear model becomes a cyclic model in which farmers and breeders work together to develop new varieties. New breeding material is developed through a collaboration of breeders and researchers. On-farm activities carried out by farmers do not conflict with institutional plant breeding. In fact, such activities are valued (or recognised) as part of the breeding process.

This approach has proven valuable for the development of plant material in marginal Third World regions. The approach can also be applied in sustainable production systems, especially in organic agriculture, in the developed nations. After all, the goal of participatory plant breeding is to develop varieties in local circumstances so that plants are adapted specifically to the prevailing growing conditions. Similar experiments are already being conducted in organic plant breeding.

Organic farming does not oppose specialisation, provided that interactions then occur at higher levels, for example by linking farms in collaborative projects. However, if a farmer is not specialised in breeding and propagation and prefers to leave part of the plant-seed interaction to someone else, he will have to ensure that he develops a close business relationship with the plant breeder/propagator. In this manner, knowledge and experience are mutually exchanged and put to use in the most optimal manner. However, in the absence of a direct farmer/breeder relationship, maintenance of diversity within crop varieties should help to ensure 'unconscious' selection for the farm or local environment.

Breeding Methods

Breeding methods never stand on their own; here, we have placed them in their agricultural and socio-economic context. For centuries, farmers had been responsible for their own seed. In the sixteenth century, domestic and international trade grew and prompted some farmers to propagate seed professionally.

Seed trading companies did not start breeding varieties until the 1950s, when the agricultural sector began its transformation into a mosaic of specialisation. The role of science in agriculture grew and the new knowledge supplied by science ultimately led to a greater control of farm circumstances through soil cultivation, land consolidation, chemical fertilisers and pesticides. All this required varieties to be highly uniform. In the race to maximise yield and uniformity, the large variations in landraces were dismissed as suboptimal and undesirable.

Subpopulations were extracted from the broad genetic variety of landraces and a limited number of types were selected. In this race, the limitations of breeding based solely on selection soon became evident and crossing techniques were subsequently developed.

Crossing techniques launched the breeding sector into a new, second, phase. From now on, farming and breeding were two distinct activities. Breeding was carried out by specialised firms that pursued their own economic goals. The bond between breeding and farm conditions was finally broken. New varieties were characterised by uniformity and yield.

Many of these new varieties were hybrids, which are typically highly uniform, highly productive plants because of heterosis. This degree of uniformity had never been achieved in cross-pollinators. With hybrids, the selected characteristics are not passed on to progeny. Breeders therefore also had economic motives for developing hybrid varieties. However, breeders' focus on hybrids has led to a stagnation in the breeding of seed-forming varieties, which therefore have a lower yield and quality than hybrids.

The first tissue culture methods were developed in the 1930s to increase the possibilities for crossing wild varieties with cultivars in order to transfer disease resistance qualities from the wild plant. In vitro techniques facilitated crosses between increasingly dissimilar species, thus broadening a cultivar's gene pool. Micro-propagation and in vitro selection cut down on breeding time. Currently, tissue culture techniques are used in breeding to increase variation, to simplify selection and to speed up propagation so that seeds can be marketed sooner. Tissue culture techniques have become a standard element in breeding programmes, illustrating just how far removed breeding is from farm practice.

The cleft between breeding and farming is deepened by gene technology, which enables characteristics to be incorporated at an even lower level. Biotechnology does not stand on its own, rather it is a new and supplementary option in breeding.

Table 1. Historical development of the breeding process.

Phase 1 (until the 1950s)

* identify landraces
* select subpopulations for desired qualities
* cross and backcross selected subpopulations
* select for desired qualities

Phase 2 (1950s to 1970s)

* cross and backcross elite material
* selection
* in a limited number of firms: programmes for developing parent lines from exotic material

Phase 3 (1970s to the present)

* cross and backcross elite material, parent line development partly replaced by development of 'gene constructs'
* selection

Plant Breeding Objectives

The breeder's role in the agricultural chain is to supply farmers with the best varieties. However, in practice these varieties will only prove the 'best' choice and give the promised yield in the right conditions. These conditions pertain to the standardisation of the agro-ecological system through the use of chemical fertilisers, pesticides, soil disinfection and anti-lodging substances. Chemical fertilisers in particular helped standardise the production regime by distributing nutrients uniformly in the field. Soils with different production characteristics are standardised to facilitate production maximisation in the long term. However, the generous application of fertilisers and manures makes crops susceptible to disease, which in turn stimulates pesticide use and focuses breeding efforts on resistance.

In the post-war period, efforts were concentrated on increasing agricultural production. Plant breeding, too, focused on increasing the ratio of harvestable yield to the total biomass of a cultivar as, for example, with short-straw varieties in cereals. Since the 1950s, the total production volume per farm has increased tenfold and production volume per hectare almost fivefold. This is partly attributed to plant breeding programmes, and partly to intensification and standardisation in farm management. One could call it a synergy between breeding and farm management aimed at boosting production.

In conventional farming and breeding circles, the harvestable part of the plant reflects the plant's worth. But seeing a plant only as the sum of its edible parts is not a complete picture. In organic farming, crop residues are not useless waste products. This biomass may play an important role in the plant's ability to reduce weeds (plant structure, tillering speed and range, foliage, stress tolerance and hardiness), in contributing to soil condition or it may be used as straw for livestock. In other words, organic farming values non-harvestable parts of the plant more than does conventional farming, which has had to resort to chemicals to fill in for the plant where certain qualities have been selected out (for example, chemical weedkillers are used in crops where dense foliage has been selected out).

Resistance breeding is also aimed at increasing yield. Commercial breeders focus their efforts on absolute, monogenetic resistance since this can be selected fairly quickly and easily and gives the plant maximum protection against a specific pathogen. Absolute resistance does not alter the crop, thereby securing yield (and the appearance of edible parts of the crop). At the same time, however, the selection process in pathogens will be directly solely at overcoming that monogenetic resistance. Absolute monogenetic resistances are invariably overcome within a few years and new resistances must be incorporated.

For this reason, organic agriculture prefers increasing polygenetic resistance to disease and maintaining variation in resistance within the crop population. Plants remain subject to a marginal, acceptable level of disease, but the resistance tends to be more durable. Crop yield may be lower and the appearance of the edible parts of the plant may be imperfect. However, the main disadvantage is that this approach can be a difficult and lengthy process. In short, striving for speed and efficiency in breeding may conflict with finding durable, organic solutions to resistance problems in agriculture.

As stated, conventional, commercial breeders prefer absolute monogenetic resistances because of 'breeding economics'. Breeders try to keep the costs of developing new varieties down by reducing the breeding cycle. Compared to monogenetic resistance breeding, breeding for polygenetic and/or partial resistance considerably lengthens the breeding cycle. These resistances are therefore less interesting from an economic point of view. However, fundamental research programmes on polygenetic resistance are being carried out.

To minimise the duration of the breeding cycle, the basic structure of a variety is fixed. In effect, high yield varieties and elite lines are intercrossed so that a new variety can be selected in the shortest possible time. When the breeding materials do not possess the desired quality themselves, genetic variation is introduced 'from outside', for example through parent lines developed from wild plants or gene constructs.

Diversity in the breeding cycle is usually restricted to elite material. In other words: previously selected breeding material is used as much as possible. This reduces the gene pool for breeding. Landraces and wild plants are not incorporated in the breeding cycle as much as they could be because they extend the cycle, thus increasing the costs of the programme.

Cost considerations also play an important role in the implementation of biotechnology. It is generally assumed that incorporating a single characteristic into an existing variety by means of a gene construct would shorten the 'improvement cycle', compared to crossing and generative selection, while at the same time the gene pool is expanded. Using gene constructs, resistances could be incorporated that would not be possible through classical methods.

As gene constructs can be used to produce new varieties, biotechnology as a whole can be said to complement traditional breeding methods. A major disadvantage with the use of gene constructs, however, is that they are based particularly on a monogenetic approach to plant characteristics. Breeding economics also limits diversity. Profit maximisation demands a maximum distribution range. Investments need to be paid back faster because varieties have shorter life cycles.

These aspects conflict with the development of regionally adapted varieties. The greater the geographical range in which varieties are marketed, the more generic, or average, the variety's qualities must be. These varieties can be grown in many different regions in standardised conditions, however standardisation in turn reduces the specific characteristics for a region. The bottom line is that regional ecological characteristics are adapted to 'generic' seed.

Current Breeding methods for Organic Agriculture

The two basic principles of breeding are:

1) to create and collect genetic variation, followed by
2) selection of the most promising genotypes.

Genotype preferences are determined by agricultural need and by the socio-economic motives of the breeding firm. A plant's phenotype is the result of genotype - environment interaction, therefore selection at the crop level is the same as selecting for phenotype. The course of the selection process can be influenced by carrying out selections in specific growing circumstances.

Various methods exist to select individual plants from plant populations which possess the desired combination of characteristics. In mass selection, what appear to be the best plants are selected and maintained in bulk without testing the progeny separately. Plants with undesirable characteristics are eliminated. Mass selection is commonly applied in the early stages of a breeding programme when there are too few potentially desirable genotypes for repeated testing. In pedigree selection, seed-propagating plants are retained or eliminated on the basis of the performance of their progeny.

This selection method works well for self-pollinators. In cross-pollinated populations, some cross- pollination in the selection field is inevitable, making pedigree selection less useful for these types of plants. In indirect selection, selection is not based directly on a desired characteristic, for example because this characteristic cannot be observed until the plant has flowered; rather, plants are selected for a strongly correlated characteristic. A lot of research goes into finding such correlations.

The moment and focus of selection are shifting. Field selection is shifting to later generations because specific characteristics are first selected in the laboratory. Secondly, the focus of selection itself is shifting towards increasingly specific, distinctly quantitative characteristics (size, colour, blemishes etc.) which can be entered as data into the computer. This simplifies the selection process and cuts costs. However, in this process selectors lose their feeling for the plant as an interconnected whole, and reliance on breeders' special, artisan skills, the unique ability to pick out an unusual or harmonious plant, is reduced. Conventional breeders admit that this is the other side of progress and a reduction of their art. Organic breeders should be given ample opportunity to develop that earlier expertise.

All selection methods in the (experimental) field at crop level pertain to a genotype -environment interaction. Therefore these techniques are well-suited for organic farming, which is aimed at optimising the plant-environment interaction. By selecting at different locations in organic conditions, breeding can boost a plant's adaptability to regional, natural circumstances. Organic field conditions further aid the selection process. This applies to the pathogenic stress, root formation and adaptation to organic fertilisation and soil quality. These selection methods are also suitable for participatory plant breeding, involving farmers in regional selection processes by setting up experimental and selection fields on their farms.

Research on genotype - environment interactions plays an increasingly important role in population genetics. New ecological insights obtained through fundamental research will facilitate the further development of variation and selection methods at the crop level. This is likely to provide a new impulse to the breeding of seed-forming varieties, thus reducing or even eliminating the gap with hybrid varieties.

In plant breeding, crossing is the oldest and most common method of obtaining a new variety when selection within existing varieties has ceased to yield the desired result. To obtain plants with new combinations of characteristics, pollen from selected paternal plants is applied to the pistil of maternal plants. In combination breeding, these plants will be existing varieties within a species and crossing rarely causes a problem. In species-crossing, cultivars are crossed with their wild equivalent, which can sometimes only be achieved after bridge-crossing. Generally, next to the desirable traits, many undesirable wild characteristics are also transferred which are subsequently eliminated by repeated backcrossing with the cultivar. During the backcrossing process, selection occurs for the desired characteristic and undesirable characteristics are rogued out.

Hybridisation

The techniques used in hybridisation have been debated for some years now in organic circles in general and in biodynamic circles in particular. A special section is therefore set aside for the discussion of hybrids. Discussions generally focus on two aspects of hybrids: a) how natural is hybridisation and b) what are the socio-economic consequences.

In cross-pollinating crops it is particularly tricky to select a new variety; variation among plants of one variety is inevitable. Just how much of a problem this variation poses for farmers or breeders is debatable. Hybridisation is a way of achieving highly uniform varieties, if that is an absolute requirement. Parent lines have to be artificially inbred for a number of generations to achieve homozygous lines. The F1 (the hybrid variety) is the result of a cross between two homozygous inbred lines. F1 hybrids have a uniformly high vigour (heterosis). Artificially inbred lines are often weak and have little vigour. Breeders select for vigour, so that the strongest lines are used to produce the most vigorous F1. The F1 line represents the new variety, therefore large numbers of seed need to be produced.

In plants such as maize, where male and female inflorescence are physically separated, the male inflorescence may be removed mechanically. In other crops, natural male sterility is introduced in the maternal line through crossing or protoplast fusion. This technique is applied in the development of parent lines in headed cabbage. Male-fertile restorer lines (so-called B lines), whose genetic make-up is identical to that of the maternal line, are used to maintain the maternal line. In seed-forming plants, the restorer genes are also present in the paternal line, so that the F1 also has fertile pollen, in other words, it retains its ability to form seed.

The heterosis effect is sometimes seen in F1 hybrids resulting from self-pollinators (eg lettuce, tomato, wheat), but for practical reasons few such hybrids have yet been produced. In these cases, the production of F1 plants poses a problem. It is difficult to eliminate the pollen of the female parents. In addition, the form of the flower rarely allows efficient pollination with pollen from male parents carried by the wind and/or insects. For some crops, such as tomatoes, mechanical emasculation and pollination of the female line is worth the effort because of the sheer numbers of seed produced. This seed often has a fair economic worth as well.

In general, inbred lines in cross-pollinators have less vigour. Sometimes tissue culture techniques must be used to maintain them. If we state that organic breeding methods may not affect the reproductive ability of the plant, then hybrids based on such inbred lines, such as leek hybrids, cannot be used in an organic farming system. This also applies to cytoplasmic male-sterile female lines for which no restorer line is available. These practices impede a sustainable use of the variety in cultivation.

If cytoplasmic male sterility is applied, the female line, and in seed-propagating plants the F1 generation, can be maintained with a restorer line. This illustrates a degree of specialisation at crop level, with respect to reproduction, which may be considered acceptable for organic agriculture.

Socio-economic consequences

Organic agriculture has largely gone with the flow of leaving propagation to specialised companies, mostly by choice. Hybrid varieties can be suitable for one-time propagation in organic conditions, but do not allow further adaptation in subsequent generations according to a cycle of propagation and selection. However, there is no question of choice when only hybrid varieties are available for a certain crop. This restriction is felt most by farmers who could feasibly fit the propagation-selection cycle into their farm management system, such as with cereals. Thus hybrids have no place in a participatory breeding system.

Hybrids prevent further propagation not only by farmers but also by breeders from rival firms. Unlike with seed-forming varieties, breeders will not part with their hybrid parent lines, thus preventing the use of these lines in other breeders' programmes. The result is erosion of the genetic base of a cultivar. For many crops, breeders' interest in seed-forming varieties has waned and the lack of improvement is reflected in the price of these varieties. In many crops, such as maize, cabbage and tomatoes, seed-forming varieties have been wholly replaced by (more expensive) hybrids.

It is generally believed in organic circles, that many seed-forming varieties would perform just as well as today's hybrids if the same amount of capital had been pumped into their development. For the reasons mentioned above, hybrids are not really compatible with organic farming principles. However, since not all hybrids are alike, they should not be judged as a group. Hybrids therefore fall in the grey category of what is and what is not acceptable in organic agricultural practice.

The agricultural advantages of hybrids are already so embedded in organic agriculture that we propose allowing some hybrids, which are more or less in keeping with organic principles (the 'yes, provided that' approach). At the same time, the organic sector calls for an initiative to develop seed-forming varieties to prevent genetic erosion. The following propositions can be formulated:

Breeding at the cell level

Breeding at the cell level comprises tissue culture. Tissue culture techniques fall under the general term 'biotechnology' and in many cases have been used in breeding for many years. These techniques have paved the way for genetic engineering. Until now, organic agriculture has refrained from judging breeding techniques at the cell level. However, if we are to draw up preconditions for a breeding system according to organic principles, these techniques, too, should be assessed for their suitability for such a system.

Tissue culture methods are based on the premise that each cell in a plant can potentially grow on to form a whole new plant. This process is called regeneration. The entire operation is carried out in sterile conditions on a nutritive medium of water, minerals, sugars and vitamins. The addition of synthetic plant hormones (auxin and cytokinin), which balance cell division and differentiation, helps direct the regeneration process along the desired lines. The plants are cultured in a climate-controlled room.

Light intensity, length of day and daytime/nighttime temperature can be set and controlled.

Tissue culture methods can be divided into three general groups:

a) methods to induce genetic variation. For example in vitro pollination and embryo culture play an important role in species crosses, it increases the gene pool of the cultivated crop;

b) methods to accelerate the analysis and selection of large numbers of seedlings by preliminary selection on the culture medium;

c) methods to obtain large quantities of plant material in a short space of time.

Up to a point, organic agriculture has no objections to these techniques because, from an organic viewpoint, the cell is the most basic form of life. This life-form remains intact during the tissue culture period. However, if organic breeding takes the road of developing varieties that are optimally adapted to organic farming conditions, organic breeders will be taking a detour with tissue culture methods, since the necessary adaptation to organic conditions still has to take place although at a later stage.

The plants are temporarily removed from their natural (organic) environment and after culture, need time to adapt again to organic growing conditions. Moreover, lines which may have possessed favourable characteristics for organic farming may be lost during the mutation induction and selection steps of the culture process. When crossing takes place on the plant and selection occurs in the field, breeders can select plants that perform well in organic conditions right from the start. When tissue culture methods are used, this selection must occur at a later stage as the cultured plants adapt to their new environment.

Cereal Breeding

In Germany and Switzerland several small cereal breeding initiatives have been launched under different regional conditions. A group of biodynamic cereal breeders (Heyden, Irion, Kunz, Spiess) annually compares notes on their progress with lines and varieties at various sites.

Among them is Peter Kunz, a Swiss breeder who has been breeding cereals for biodynamic farming for over 15 years. In 1996 he came up with his first spelt variety, Alkor, which has now been released on the Swiss market. Applications for three other varieties will be made soon. Spelt, a cereal that is closely related to wheat, is a robust crop and gives high nutritional value at relatively low yields. As the basis for his cross- breeding work, Kunz collected 2500 varieties from various gene banks to get the most suitable varieties for organic conditions. Kunz is also breeding new wheat varieties. He made an in-depth study of the origins, developments and characteristics of wheat, based on the relevant literature and an extensive screening of 500 old and modern wheat varieties. This forms the basis of his investigation into a 'fitting' shape for wheat.

'What we aim for', he writes, 'is breeding a variety that gives the most characteristic feature of wheat its due in a well-balanced whole. On the basis of our phenomenological study we gave preference to types where the vegetative and generative elements were clearly distinguished both in the plant's development and its morphology. You get plants with long top internodes which can easily be observed in the field. This criterion leads, almost as a side-effect, to a lesser susceptibility to ear fungi, the ear being in a drier micro-climate.'

Specific selection criteria are baking quality, contribution to weed restriction and resistance to smut. To develop his selection criteria he also carried out comparative studies on baking quality. On account of the great variation in altitude, soil and climate in Switzerland, Kunz carries out his work on farms in different regions (Standort-orientierte Zùchtung).

Vegetable breeding

In Germany, Thomas Heinze co-ordinates an association for the propagation and breeding of vegetable seeds. It organises two meetings a year for its 70 biodynamic members to compare notes and to match training sessions and practical arrangements. Each breeder is in charge of one or more varieties. Forty-eight vegetable species and 158 varieties are now available for public sale. Their range comprises varieties bred under the license of other (conventional) breeders, their own selections, and a number of very useful, free varieties which can be grown again on account of the intensive conservation of selected resources. The seed stock is collected at Bingenheim, where it is tested, packaged and marketed. Important breeders here are Thomas Heinze, Dieter Bauer, Amadeus Zschunke and Ulrike Behrendt (Behrendt, 1983). They aim at varieties that do well in organic conditions, varieties with a well-developed root system, good adaptability and resistance to fungal diseases and insect pests.

In the Netherlands, Jan Velema and Taco van der Made (Vitalis Biologische Zaden BV) have bred vegetable varieties for commercial organic growers since 1994. They have recently started crossing and selecting on a modest scale. Velema notes that breeding aims for lettuce in organic conditions are very similar to those of conventional programmes, but priorities differ. By selecting and propagating under organic conditions, other characteristics are accentuated in existing varieties than would be for conventional farming. The breeding process then continues using these plants.

Velema is working on a cauliflower hybrid programme based on self-incompatibility. The difference in vigour between stable varieties and hybrids proves greater under organic than under conventional conditions. This is because hybrids develop better root systems under organic conditions.

Jan Velema also works with pumpkins. Conventional pumpkin varieties generally come from the US and Japan since there is little pumpkin breeding in Europe. Vitalis has taken it up and breeds varieties that meet European standards. Pumpkins are becoming very popular with organic growers. Vitalis also breeds a number of other crops on a small scale.

Chris and Harry Douwes run De Bolster, where they propagate organic seed for small growers and gardeners. They have a wide selection of vegetable and flower seed (250 varieties in all). The best plants are selected to produce seed for the next year. The firm aims to grow strong varieties based on years of growing and selection. Seed stock for new species and varieties is collected from other breeders, botanical gardens, gene banks and private growers.

Environmental Aspects

Selection for organic breeding is deliberately done in regional, organic conditions to allow natural circumstances to play a part. Some biodynamic growers believe there is no need for cross breeding; that new varieties can be bred from the variation within existing varieties. One such method which is still being developed is the ear bed method where selection is based on a grain's place in the ear. Variation can also be developed from growing and selecting successive generations of one variety in vastly different environments. This method is still being developed and its usefulness has not yet been tested.

The interaction of different environments is being investigated and so is the interaction of sowing dates. In some cross pollinating winter cereals such as rye it is found that early sowing dates lead to improved baking quality; later sowing dates give better quality seed.

5

Bioactive Protectants

For small farmers, the main purpose in storing grain is to ensure household food supplies. Farm storage also provides a form of saving to cover future cash needs through sale or for barter or gift-exchange. Small quantities of grain are also stored for seed. Farmers who produce a surplus may also store grain for sale later to take advantage of seasonal price rises.

Farm storage systems must provide maximum protection against deterioration of the commodity by inclement weather and pests, and also to deter theft. Traditional farm storage systems have evolved over long periods to satisfy these requirements. Most are well adapted to their environment and losses are generally low, often below 5 percent of grain weight over a storage season. However, for resource poor farmers living at or near subsistence even losses of this magnitude have important implications for food security. Rectifying these losses can only be achieved by subsistence farmers if changes are made to the traditional system of storage which bear no cost (other than of the farmers own labour), such as improving the design of the storage structure and using grain protectants which occur naturally in the local environment.

Higher losses may occur if the equilibrium of the traditional post-harvest system is disrupted. New, introduced crop varieties that do not possess the good storage characteristics of the traditional varieties may be more susceptible to pest attack; for example, storage losses in hybrid maize due to insect pests may exceed 25 percent by weight. These newer high-yielding varieties, grown by more progressive small-scale farmers, may produce grain in quantities, which exceed the traditional storage capacity. In the past, these farmers have depended upon parastatal grain marketing organizations and grain traders to take the surplus soon after harvest.

The introduction of grain market liberalisation in many countries has resulted in a rapid decline or elimination of marketing organisations and the emergence of a more important private grain trade. Farmers now have an opportunity to take advantage of seasonal price rises but the benefits can only be achieved if grain is held longer on the farm with no deterioration in quality. Thus the farmer now faces new problems associated with conserving grain safely on the farm and he/she needs advice on the most cost-effective type of storage system and methods of pest control to use.

Insect Control Methods

Traditionally, protectants against insect infestation fall into two groups: those materials such as ashes, minerals and oil, in which physical barrier effects are responsible for the control of insects; and the use of whole plants, or parts of plants where there may be some chemical insecticidal or repellent effect.

Physical Control

The use of chemically inert materials such ashes, sand or other minerals, powders or seeds in large quantities to fill up the interstitial space in grain bulks and provide a barrier to insect movement is quite widespread. For example; farmers living on the Lilongwe plain and in the Dedza hills of Central Region in Malawi commonly use ash from the cooking fire to protect beans stored at the home for both consumption and seed; in India sand is mixed with both cereals and pulses for storage; and in Indonesia peppers, beans and sesame seed are mixed with ashes or sand. The abrasive nature of such materials may also help control infestation by damaging the insect cuticle leading to dehydration and death.

Recently, silica-based inert materials (such as silica aerogels and diatomaceous earths), have proven to be very effective in field trials under certain climatic conditions, although at present the cost and lack of availability prevent their use at the farm-level in developing countries. It has also been proposed that such materials may be used as carriers for insecticide active ingredient to increase their effectiveness under a range of climatic conditions. However, silica-based dusts have not found universal acceptance, firstly, because of the potential respiratory health hazards which may be caused by inhalation of their finely divided particles and, secondly, because of the possible damage to grain handling machinery which might result from the abrasive properties of these materials.

It is now well established that vegetable oils are very effective in controlling certain species of bruchids by their ovicidal effect. The main concerns over the use of oils however is the availability, price and potential problems with taint as off-flavours may be imparted to the grain.

Other physical methods of insect control which could be employed at the farm level are thermal disinfestation, the commodity being spread out in a thin layer on the ground or on a platform and is exposed to the sun for several days; and hermetic storage, the commodity being retained in sealed, air-tight containers such as clay pots or metal bins. Thermal disinfestation is suitable for small quantities of produce and for areas which have extended periods of hot, dry weather but protection is only afforded for a few months as the technique does not provide the grain with residual protection; the insect pests tend to be driven away from the grain by the heat but not killed.

Hermetic storage is suitable for protecting small quantities of seed material but is generally too expensive for use with food grain. The use of traditional underground pits is widespread in some areas in which the physical environment is conducive to this type of storage, i.e. hot and dry with very low water tables, such as occur in Ethiopia, the Sudan and Somalia.

Chemical Control

The use of the fumigant gas aluminium phosphide (phosphine) at the farm level is becoming increasingly more common where it is available to farmers, particularly in the Indian sub-continent and West Africa. This trend is a matter of some concern as farmers are rarely sufficiently trained in the safe handling and use of such products and its misuse is likely to lead to accidental poisoning and the development of insect resistance. Its use is to be discouraged where possible although if its use continues to spread it may be necessary to ensure that extension workers are sufficiently knowledgeable and trained in the application of the fumigant to be able to train farmers in its safe use and handling.

Conventional insecticides, as dilute dust formulations which are admixed with the commodity, are safe and effective, providing that only those chemicals that have a label recommendation for use on stored grains are used (such as pirimiphos-methyl [Actellic: Zeneca Agrochemicals], methacrifos [Damfin: Ciba Geigy] and fenitrothion [e.g. Sumithion: Sumitomo]), and proper attention is paid toward user safety aspects. In many developing countries availability of suitable products is poor, and often inappropriate, dangerous chemicals, such as fenthion, lindane and DDT, may be used instead. The direct application of a pesticide to a food commodity, known as 'admixture', will always produce a residue which will be more or less persistent depending upon the chemical nature of the pesticide used (and climatic conditions and, to a lesser extent, the nature of the commodity to be protected). This may however create a potential hazard or, at least, a source of possible anxiety to the user of the commodity.

Application of chemicals to surfaces of store walls and floor by spraying with wettable powder or emulsifiable concentrate formulations to produce residual surface treatments are not recommended due to the very short effective persistence, often less than one day under tropical climatic conditions. Such residues will often contribute to the development of insect resistance, which occurs when insects are exposed to sub-lethal concentrations.

Insecticides used in places where foodstuffs are stored or handled, especially when used for admixture with commodities, should always be chosen from those that have international and local approval for the particular purpose, and established maximum residue limits that allow effective application rates. Most developing countries have very limited legislation controlling the use of chemicals and instead rely upon recommendations made by United Nations Organizations to identify acceptable compounds which are safe to handle and consume.

International regulation pertaining to the use of insecticides on food stuffs is governed by the United Nations Codex Alimentarius Commission (CODEX). Codex specifies chemicals and residue limits, which are permissible in traded grains and other food commodities. Its specifications are based upon recommendations provided by the FAO and WHO Joint Meeting on Pesticide Residues (JMPR). This body accumulates scientific evidence concerning insecticide efficacy and toxicity and recommends maximum permissible residues for different food commodities. These maximum residue limits (MRL), expressed as quantities of active ingredient (ai) of the chemical, take into account the acceptable daily intake for human beings of each compound.

Only a small number of the insecticides marketed for the protection of agricultural produce have received clearance by Codex and have MRLs recommended for durable food crops. Some of those that have approval include the pyrethroids: permethrin (2 mg/kg on raw cereals), deltamethrin (1 mg/kg), bioresmethrin (5 mg/kg), phenothrin (2 mg/kg) and fenvalerate (2 mg/kg); and organophosphorus compounds: malathion (8 mg/kg), pirimiphos-methyl (10 mg/kg), fenitrothion (10 mg/kg), bromophos (10 mg/kg) and chlorpyrifos-methyl (10 mg/kg).

Organophosphorus compounds (organophosphates) are the most widely used group for storage protection. In the absence of insect resistance, which may affect the choice of an insecticide, the choice between various organophosphates relates mainly to marginal differences in toxicity and specificity. In general, pyrethrum and the synthetic pyrethroids are more expensive than the organophosphates, but their relative efficacy against the bostrichid beetles R. dominica and P. truncatus makes them cost-effective where these pests are a problem; pyrethrum, a plant extract, is however photolabile and quickly becomes ineffective in daylight.

The control of insect infestation in maize stored on the cob also presents special problems. Unless the sheathing leaves are removed, any insecticidal spray or dust may fail to reach the insects on the grains themselves. Some control of infestation is possible even with the sheath intact with insecticides like pirimiphos-methyl, but the control achieved will depend upon many factors including grain moisture content, ambient humidity, the condition of the cob sheath, the type of storage structure used and the pattern of infestation. In general, the best solution is to dry the cobs as quickly as possible and then shell the grain so that an admixture treatment can be applied more effectively. This is certainly the best treatment where P. truncatus is a serious pest , provided that natural drying in the crib does not take more than one to two months and that initial infestation levels are not exceptionally high.

Plant Materials

The wide-scale commercial use of plant extracts as insecticides began in the 1850s with the introduction of nicotine from Nicotiana tabacum, rotenone from Lonchocarpus sp, derris dust from Derris elliptica and pyrethrum from the flower heads of Chrysanthemum cinerariaefolium. Both D. eliptica and Lonchocarpus belong to the Fabaceae family and have been found to contain 3-20 percent rotenone and also rotenoids, the chemical components which show insecticidal activity. C. cinerariaefolium belongs to the family Asteraceae and the flower heads contain six identified active compounds, pyrethrins I and II, cinerins I and II and also jasmolins I and II.

Several other traditionally used plant preparations have found local commercial markets, for example Ryania speciosa (Ryania) (Flacourtiaceae) which contains an insecticidal alkaloid, and Haplophyton spp. (Apocynaceae) which have been used in the West Indies and Mexico for crop protection. Nicotine, an alkaloid from some Nicotiana species (Solanaceae) and the related compound anabasine, from Anabasis species continue to be used against various insect pests in spite of being highly toxic to mammals. Constituents of many aromatic plants used for flavouring or medicinal purposes have been found to possess insecticidal properties.

Recent surveys of desert and semi-desert plants have revealed a range of sesquiterpenes, benzopyrans, chromenes and prenylated quinones that are repellent or cytotoxic. Some plant families may accumulate a restricted number of anti-insect chemicals, so-called secondary metabolites, whilst others possess a wide variety of different structural compounds. There is renewed interest, for various reasons, in both plants and other alternative materials, such as silica-based inert dusts and insect growth regulators, for use as stored-product protectants. These include the development of insect resistance to synthetic insecticides, fears over misuse and overuse (during application) and the potential problem of insecticide residues affecting consumers, wildlife and the environment. Also, the increased cost of insecticide development has led to a reduction in the number of new pesticides being evaluated by chemical companies for use in storage applications.

In developing countries, the high cost of using imported synthetic insecticides may result in a drain on limited reserves of foreign exchange. These problems are compounded by the difficulties and expense of packaging and distributing chemicals in a form which is suitable for on farm use, requires no sophisticated equipment for application and is cheap. The use of locally available plants avoids the need to establish complex mechanisms for pesticide distribution; the community can collect or grow the plants itself. The use of plants materials for storage protection is sustainable; they can be continuously propagated year after year; biodegradable; and do not have any negative impact on the environment as long as care is taken to avoid the propagation of plants from foreign ecosystems which might, therefore, become established as weeds.

In considering the natural materials required for effective pest control it is clear that large quantities are often required for satisfactory control. In some instances this may result in some harm to the local environment; for example if large quantities of ash are required, deforestation may result. Many non-cultivated plants have been used traditionally in foods and will therefore present no risk to health.

Although commonly regarded as safe, many plants contain noxious compounds, which may render them unsafe for both animals and humans to consume. Therefore, the toxicity of plant materials, either directly admixed into foodstuffs or of extracts which are applied onto the produce, may be unacceptable and even long-standing use does not guarantee a level of safety which can be recommended unconditionally.

6

Insertional Mutagenesis

The completion of sequencing of the genome of Arabidopsis thaliana marks a major milestone in plant biology research. The determination and analysis of the genome sequence of Arabidopsis have provided the first detailed description of the genetic blue print of a higher plant and revealed several novel processes involved in plant growth and development. The availability of genome sequence of Arabidopsis is likely to broaden and accelerate further research in plant sciences.

With the availability of complete genome sequences, the focus of research world wide is now on functional genomics and to assign functions to newly identified DNA sequences. Having identified a new sequence, the comparison with sequences in the databases is the simplest way to obtain functional information. Although efficient bioinformatics tools are becoming available for the annotation of genome sequences, in silico analysis by itselfis only indicative and is generally not sufficient to define function of a gene. Even in those cases where some indications become available from in silico analyses, experimental evidences are required.

One of the most challenging tasks before plant scientists is assigning functions to a large number of plant genes. The functions of ~69% of the genes were classified according to sequence similarity to proteins of known function in all organisms. Definitive functions for individual genes have been thoroughly established for less than 10%. Out of the ~26000 genes identified in Arabidopsis, the functions of only a few thousand have been defined with great confidence and more than 30% of the predicted Arabidopsis genes could not be assigned any specific function.

A variety of approaches are used to clone and gather information about the function(s) of gene(s). Among these, insertional mutagenesis has been extensively used for cloning genes, promoters, enhancers and other regulatory sequences from Arabidopsis. The significance of insertional mutagenesis in functional genomic studies and compile the developments on cloning, characterisation and identification of genes and promoter elements in A. thaliana, using the T-DNA tagging strategy. In addition to describing the various experimental approaches, we also point out some of the common difficulties and problems associated with the methodology.

Strategies used for cloning and characterisation depend upon the information available about the gene or its product. The information, which can be exploited, includes thespatial and temporal expression patterns of a gene; for example, the presence of its mRNA and/or protein in different cell types, during development, during pathogen infection, or in different environments will have to be examined.

Expressed sequence tags (EST's) and microarray-based techniques are some of the powerful approaches in this direction. Methods utilising the tools of proteomics like SDS-PAGE, MS, MALDI-TOF, Yeast two-hybrid system etc. have proved very useful in elucidating gene function.

Gene suppression (deletion / mutation knock out) or over-expression permits the gene sequence to be linked to a phenotype from which the function of the gene can be deduced. TILLING (Targeting Induced Local Lesions In Genomes), another strategy involves traditional chemical mutagenesis followed by high throughput screening for point mutations and allows identification of allelic series of induced point mutations in genes of interest. Conversely map based cloning is another tool to identify a gene known only by its phenotype.

Gene disruption is a powerful tool for obtaining knock out mutants that helps in ascertaining biological function of the numerous uncharacterised open reading frames (ORF's), revealed by the genome sequencing project or represented in expressed sequence tags. Gene knock out systems provide a direct route to determining function. Most other approaches to gene function are correlative and do not necessarily prove a causal relationship between gene sequence and function. For example, DNA chips provide an exciting means to discover conditions under which gene expression is regulated on a genome wide scale. However, because factors other than mRNA level also determine the activity of a gene product in situ, expression studies even when done on a genome wide scale cannot prove a causal relationship.

By contrast, the availability of a null mutation for the gene of interest allows one to directly monitor the effect of this deficiency on the organism's ability to function. Whereas gene replacement through homologous recombination is now a routine in yeast and mice, in plants exploiting homologues recombination for targeted insertion has not been easy. Although Terada et al. describe a strategy to achieve homologous recombination in rice, the procedure needs further refinement for routine application in plants.

Antisense or over expression is another strategy, to identify gene function. However, these are laborious approaches and results are often inconclusive or impossible to interpret. Recent reports of successful use of double stranded RNA mediated interference (RNAi) approach in C. elegans, Arabidopsis is certain to open new vistas for sequence- specific inhibition of gene function in Arabidopsis.

Mutational approaches have been successfully usedfor the study of genetic and molecular bases for many traits in plant biology. Access to the mutation is obtained using positional cloning strategies. This strategy is facilitated in model species such as Arabidopsis, for which dense genetic maps with many visible and molecular genetic markers exist, and for which complete physical map consisting of a collection of overlapping cloned DNA fragments and total genome sequence are already available. The limiting factors for this approach are the time and effort required for creating the mapping population and the fine mapping of the mutant locus.

T-DNA Tagging

The use of insertional mutagenesis in principle provides a more rapid way to clone a mutated gene. DNA elements that are able to insert at random within chromosomes such as transposons or the T-DNA of Agrobacterium tumefacians, can be used as mutagens to create loss of function mutations in plants. Since the sequence of the inserted element is known, the gene in which the insertion has occurred can be recovered, using various cloning or PCR-based strategies. An advantage of using T-DNA as the insertional mutagen as compared to transposons is that the T-DNA insertions do not transpose subsequent to insertion and are chemically and physically stable through multiple generations.

The T-DNA not only disrupts the expression of the gene into whichit is inserted, but alsoacts as a marker for subsequent identification of the mutation. Since Arabidopsis introns are small and there is very little intergenic material, the insertion of a piece of T-DNA on the order of 5 to 25 kb in length generally produces a disruption of gene function. If a large enough population of T-DNA transformed lines is generated, there are reasonably good chances of finding a transgenic plant carrying a T-DNA insert within any gene of interest. Mutations that are homozygous lethal can also be obtained and maintained in a population in the form ofheterozygous plants. Low copy number and random nature of insertions are considered to be the advantages of T-DNA induced mutagenesis approach.

Extensive work in T-DNA tagging of Arabidopsis has become possible because of improvements in techniques for Agrobacterium mediated transformation. The original root explant method of Valvekens et al. allowed one to isolate a few transformed plants, via a laborious tissue culture process. Tens of thousands of transformed plants were beyond reach, until Feldmann and Marks devised a method for producing independent T-DNA transgenic lines via seed transformation. The development of transformation methods based on dipping whole plants into Agrobacterium suspensions has made it possible to generate hundreds of thousands of insertional mutations necessary for saturation of the genome while minimising the effect of somaclonal variation associated with the process of in vitro culture and regeneration.

The advantage of T-DNA over classical mutagens is that the Arabidopsis sequences flanking the insertion site can be isolated easily. This simplifies the identification of genes corresponding to interesting mutants. When screening populations that have been mutagenised with an insertion element, not all of the new phenotypes will be due to insertion mutations. It is important to establish linkage between the mutant phenotype and the insertion element. Even if some linkage can be established, this does not exclude the possibility that the phenotype is caused by a nearby but independent mutation.

Consequently it is necessary to identify the gene disrupted by the insertion and test for complementation of the mutant phenotype by the wild type allele in transgenic plants. Because it is not difficult to identify sequences flanking insertion sites, it is reasonable to dispense with co-segregation testing, and proceed directly to identify the gene interrupted by the insertion, followed by a test for complementation of the mutant phenotype by the wild-type version of the gene. Such complementation data provide definitive proof that the phenotype of interest is in fact caused by the insertion mutation. Another way of confirming the same would be by characterising an independent mutation in the same gene displaying an identical phenotype. Allelism tests can be used with any recessive mutants exhibiting the same phenotype.

A variety of strategies have been devised to generate and isolate mutants in known genes of Arabidopsis by T-DNA or transposon insertional mutagenesis. In these methods, large populations of tagged mutants are generated, which can then be screened for insertions in specific genes. Alternatively, the insertion tags can be individually sequenced and compiled in databases that can be searched for a gene disruption event of interest. This is becoming easier now as genomic sequences of many insertion sites are becoming available.

A highly efficient procedure for obtaining mutants in genes identified in sequencing programs takes advantage of the availability of large collections of plants mutagenised by an insertion element. This procedure makes use of the specificity and sensitivity of the PCR reaction to screen for insertions within regions of interest in a large population of mutagenised plant lines. Using oligonucleotide primers from the insertion element and from the gene of interest, it is possible to detect an insertion event within the gene, even in case of a pooled DNA sample. The sensitivity of the PCR is so high that it is possible to detect such an event in large pools (up to a few thousand) of mutagenised plants. The pool is then repeatedly subdivided until a single plant carrying the desired insertion is identified. The identified mutant plants are then tested for phenotypes that are predicted to result from loss of function of the gene. This is becoming easier now as genomic sequences of many insertion sites are becoming available.

Winkler et al. have developed an efficient reverse-genetics protocol that uses expedient pooling and hybridisation strategies to identifyindividual transfer-DNA insertion lines. They have screened a collection of 6000independently transformed lines using this protocol to systematically isolateArabidopsis lines containing insertional mutations in individualcytochrome P450 genes.

After segregation analysis of a large number of transformants, concluded that the average number of independent inserts is 1.5 per diploid genome with 57% of the transformants containing a single insert and 25% containing two. Also within the context of the target gene, insertion appeared to be random, as T-DNA inserts were found in exons, introns, and in the 5′ and 3′ flanking regions of the more than 30 genes that were characterised. In addition, after screening of the first 14000 transformants no obvious hotspots for insertion of T-DNA were detected. However, genome wide analysis of the distribution of integration sites conducted by Alonso et al. on a T-DNA mutagenised population of Arabidopsis revealed the existence of a large integration site bias at both chromosome and gene levels.

At the gene level a significant bias was observed against integration events in exons and introns, in favour of 5′UTR, 3′ UTRs and promoters. A similar observation was made by the German plant genomics research programme GABI-KAT. The GABI-KAT population of T-DNA mutagenised Arabidopsis thaliana lines with sequence-characterised insertion sites is being used extensively for efficient progress in plant functional genomics. After PCR-based amplification of DNA fragments spanning insertion site borders and sequencing, the data were placed in a flanking sequence tag (FST) database describing which mutant allele was present in which line. Analysis of the distribution of T-DNA insertions revealed a clear bias towards intergenic regions. Insertion sites appeared more frequent in regions in front of the ATG and after stop codons of predicted genes.

Genome Resources for Public Use

Large collections of the T-DNA insertion lines are being provided to the Arabidopsis stock centres at Ohio State University (USA) and Nottingham (UK) by the individuals/groups who constructed them. More than 175,000 T-DNA insertion lines of various types are already available from the Arabidopsis Biological Resource Center (ABRC). ABRC sends the stock free of charge to any laboratory that cannot afford the associated charge. The stock centres have a distribution agreement.

According to which, North and South American users are expected to order seed stocks from ABRC, and European users are expected to order seed stocks through NASC. The laboratories in other locations may order seed stocks from either centre. A number of lines require the signing of a material transfer agreement (MTA). For obtaining GABI-Kat T-DNA lines the costs is • 499/- per line for academic institutions and are available only for research purposes. However, for profit institutions a contract needs to be set up between the company and Garching Innovation (the MPG technology transfer organisation).

Furthermore several groups are working to isolate and sequence DNA fragments flanking insertion sites and are depositing sequences in databases. When these databases reach sufficient size, it will be, possible to find an insertion in any gene of interest simply by consulting the databases and ordering seed from the stock centre. Systematic sequencing of insertion sites in various populations have been initiated by several group.

Table 1. Functionally characterised genes from Arabidopsis using T-DNA.
a. Forward genetic approach.

S. No.	Genes	Functions
1	GL1	Trichome development
2	CPR-20	Disease resistance signalling
3	GI	Phytochrome signalling
4	MYB75	Anthocyanin production
5	LEP	Leaf development
6	DPE1	Starch breakdown
7	Dfl1	Auxin control of root development
8	KAN	Organ polarity
9	BRS1	Brassinosteroid signaling
10	QUA1	Biosynthesis of pectins
11	ATE1	Leaf senescence
12	PRPS17	Leaf senescence
13	PLS	Auxin-cytokinin homeostasis to modulate root growth and leaf vascular patterning

14	CPL1 and CPL3	Abiotic stress signalling growth and development
15	OSM1 and SYP61	Stress tolerance and ABA regulation of stomatal response
16	HCF152	RNA processing
17	EXO	Cell division
18	OMT1	Lignin and synapoyl ester biosynthesis
19	ULI3	UVB light response
20	TPT	Acclimation of photosynthesis to environment
21	AMS	Tapetal and Microspore development
22	BXL1	cell wall metabolism
23	Cpn60 beta	Molecular chaperonin

b. Reverse genetic approach

S. No.	Genes	Functions
1	PGP1	Photosynthesis
2	SEC and SPY	Gametogenesis and embryogenesis
3	PGP1	Photosynthesis
4	AMT1;1	NH4+transporter
5	LACS9	Chloroplast Acyl-coenzyme A (CoA) synthetase
6	DIR1	systemic resistance signalling
7	GEM1-1 and GEM1-2,	cytokinesis
8	ECA1 Ca2+	pump
9	MRE11 and KU70	Telomere maintenance and DNA repair
10	IRT1	Metal transporter
11	AtTOP6B	DNA replication
12	MMT	Selenium volatalisation
13	RCN1	ABA signal transduction

14	CKI1	Megagametophyte development
15	MMP	Flowering and early senescence
16	AtOPT3	Embryo development
17	AtpOMT1, AtpDCT1, AtpDCT2	Oxoglutarate/Malate and Dicarboxylate Transporters
18	PEX10	Peroxisome biogenesis
19	UVH6	DNA repair (UV resistance) and growth
20	UGT73C6 and UGT78D1	Flavonol glycoside biosynthesis
21	CAP-E1	Embryogenesis, meosis and Meristem organisation
22	LUT1	Carotenoid hydroxylase
23	SCD1 and Arp2/3	Polar cell expansion
24	CLAVATA1	Organ development
25	ATHB5	Abscisic acid signal transduction
26	TDS4	Anthocyanidin synthesis
27	AtFUT1	Xyloglucan fucosilation (cell wall biosynthesis)
28	ARR15	Cytokinin mediated signal transduction
29	ATM	Meiosis and somatic response to DNA damage
30	MAF2	Vernalisation
31	Apyrases (Atapy1, Atapy2)	Pollen germination
32	GORK	Stomatal movements and transpiration
33	FATB	Fatty acid biosynthesis
34	GGAT	Photorespiration
35	ACT7	Germination and root growth
36	PIP2;2	Hydraulic conductivity of roots

37	ADA2b and GCN5	Growth, development, and gene expression

Recently 150,000 transformed plants carrying 225,000 (1.5 insertion per line) independent T-DNA integration events was generated and precise locations determined for >88,000 loci by the Salk Institute of Biological Studies. Analysis of the insertion site sequences revealed that insertional mutations had been created in ~74% of the annotated Arabidopsisgenes.

Surprisingly, although PCR based reverse genetic screens have been available to Arabidopsis researchers, for several years, relatively few informative knockouts obtained in Arabidopsis have given rise to a visible, directly scorable phenotypes. Several reports have shown that this lack of phenotypes is presumably because of partial and complete functional redundancy besides the ability of higher plants to adapt their physiology to various stresses and constraints without undergoing morphological changes, and by our inability to detect slight physiological alterations and/or weak reductions in fitness.

In order to identify conditional phenotypes, the mutants have to be tested under a wide range of environmental conditions, as plants being sessile organisms have evolved many adaptive traits that allow them to cope with changes in their biotic and abiotic environment. Many loss of function mutants are therefore expected to be conditional and revealed only by a specific combination of environmental parameters.

Using T-DNA tagging several genes have been identified and cloned. Table 1a and 1b provide a list of genes cloned in Arabidopsis using T-DNA tagging in the recent years, using forward and reverse genetic approaches, respectively.

T-DNA lines Using Specialised Vectors

Simple insertional mutagenesis, like all gene disruption approaches, has some limitations. It is difficult to identify the function of redundant genes or of genes required in early embryogenesis or gametophytic development. To overcome these limitations modified insertional elements have been developed.

Activation trap, enhancer trap and gene / promoter trap vectors are specialised versions of insertional mutagens. In addition to screens for loss of function phenotypes, gene function can also be identified based on expression patterns or gain-of-function phenotypes. These screens are particularly useful for genes that have redundant and multiple functions during development.

Activation Tagging

Enhancer and gene trap insertions identify genes on the basis of their expression pattern. In activation tagging insertion alleles are generated using T-DNA or transposons that carry either a constitutive promoter such as CaMV 35S or multimeric enhancers from the CaMV 35S promoter. Because enhancers can function in either orientation and at a considerable distance from the coding regions, they can cause transcriptional activation of nearby genes, resulting in dominant gain of function mutations (Figure 1).

Such gene activation may produce novel phenotypes that are either redundant members of a gene family or are essential for survival. Activation tagging strategy, developed by Walden et al. has been used to generate collections of morphologically diverse dominant mutants resulting in cloning of corresponding genes.

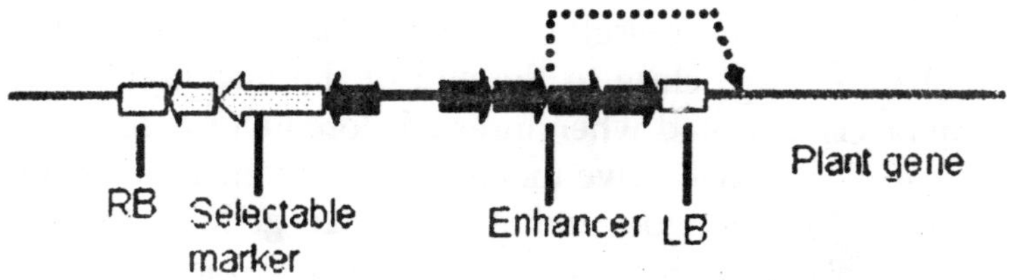

Figure 1. Activation tag vector

Majority of the gain of function mutations are on alterations of phenotypes related to plant defence responses. More recently 45000 T-DNA insertion lines were generated using an activation tag vector. Out of the 1194 T-DNA insertion sites 1010 have their tags inserted in or close to a predicted gene and nearly half of these genes were those with unknown functions. The interfered genes were distributed in all five chromosomes and were involved in a variety of processes including metabolism, transcription, plant defence, signal transduction etc.

Promoter and Enhancer Trap Vectors

With the ever increasing reports of creation of transgenic in various crop plants, the search for novel genes and variety of regulatory elements of DNA required for controlled expression of the introduced genes is gaining more importance. Regulatory elements that impart a tissue-specific, stage-specific and/or environmental-stimuli-specific expression to the transgene are being identified and cloned using T-DNA based vectors that have been designed to identify and clone such regulatory sequences.

The general principle behind this approach is to integrate a reporter gene that either lacks a promoter (gene/promoter trap) or carries only a minimal promoter (enhancer trap), at random sites in the wild type genome. A reporter gene cassette containing a minimal promoter (enhancer trap) close to the end of the insertion element can be cis activated when inserted close to a transcriptional enhancer that will drive the expression of the reporter gene (Figure 2). The gus (uid A) reporter gene is the most commonly used reporter gene system in plants, because of the absence of endogenous ß-glucuronidase (GUS) activity in most plants and the opportunity to visualise the presence of the enzyme by sensitive histochemical techniques.

Mutagenised populations can be screened for lines expressing reporter gene in specific cell types or in specific environmental conditions. Genes with interesting expression patterns and their promoters can be isolated from such lines. Small populations of promoter trap lines have been generated in our laboratory using promoter trap vectors carrying promoterless GUS gene. The T-DNA tagged lines generated are being used for identifying novel genes and tissue specific promoters. Apart from the ease of identifying redundant genes this method helps detection of insertions in UTR's. An insertion in the 3' UTR's of genes generally will not lead to suppression of gene expression as the coding region remains intact, but can lead to reporter gene expression enabling detection of the otherwise hidden mutation. Repeated observations of a significant bias towards T-DNA integration in promoters and UTR's justify a high frequency of GUS expressing lines in the mutant population.

Promoter trap lines can also be exploited to mark certain cell types for developmental studies. Enhancer traps allow an ingenious way of identifying genomic sequences

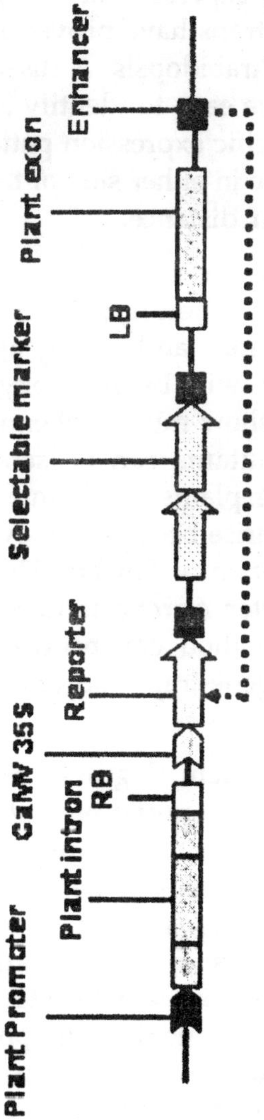

Figure 2. Enhancer trap.

expressed in precise developmental patterns. In the case of GUS as the reporter gene, the presence of GUS activity in a particular organ or cell at a particular developmental stage will identify sequences expressed at this place and time. Indeed, GUS enhancer traps have proved successful in detecting novel genes in Arabidopsis. A disadvantage, however, is that it may not be easy to identify and locate the enhancer causing the specific expression pattern, since the enhancer element could be in either side of the T-DNA and could also be at a far off distance.

Plants for mutant screens are usually grown from T2 seed because most mutant phenotypes result from homozygous recessive mutations and first generation of mutagenised plants, which will be heterozygous for induced mutations, will not show mutant phenotype. The T1 generation of T-DNA mutants can be screened for dominant mutation, as these plants are hemizygous for the insertions. Hemizygotes arise because the ovule is the target for T-DNA transformation of flowers. Thus, the T1 generation seeds obtained after Agrobacterium mediated floral dip transformation method can be used for the screening of activation tag, gene-trap, and enhancer trap lines.

In the gene/promoter system (Figure 3 and Figure 4), insertion of the promoter-less reporter not only disrupts normal gene function but also activates expression of the reporter gene. Because expression can be monitored in heterozygous plants, gene trap system is useful for studying expression patterns of most plant genes, including essential genes that cause lethal mutations. This system is convenient for observing mutant phenotypes because reporter activation indicates the location, condition and the time of expression for the disrupted gene. In Arabidopsis, activation of reporter genes has been observed in ~30% of the transgenic. Some of the recently developed gene trap systems contain an intron with multiple splicing acceptor and donor sites in each of the three reading frames in front of the coding region of the reporter gene.

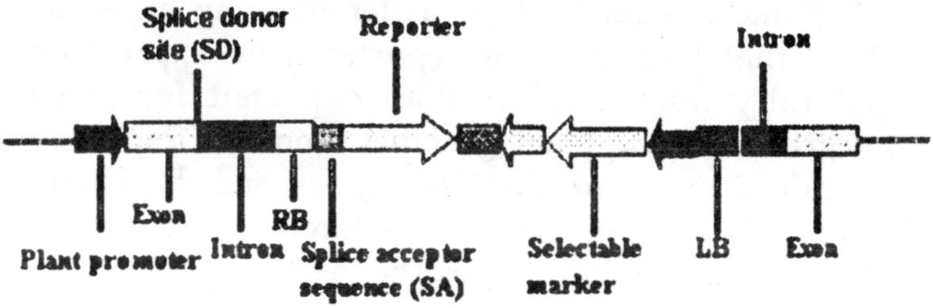

Figure 3. Gene trap vector

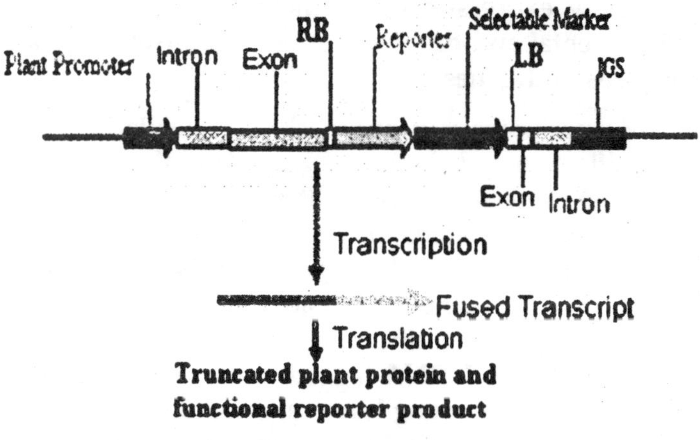

Figure 4. Prompter trap

These constructs allow reporter gene expression even when the insertion occurs in intron. As a consequence, expression is also observed at high frequencies in Arabidopsis plants transformed with such a gene trap system. Selection or screening for gene specific insertions can also be accomplished independently of promoter activity, as the expression of the marker in the gene trap vector is dependent not only on the activity of an endogenous promoter but also on transcript processing (splicing) and formation of translational fusions.

Some ten years of promoter trapping experiments show that frequency of reporter gene expression is generally much higher than expected for random integration in the plant genome. Promoter trap strategy provides a means of identifying genes, and characterizing in vivo, their expression patterns, throughout the plant life cycle in viable heterozygotes.The insertion of T-DNA into the tagged gene also facilitates its cloning. During the last few years promoter trap lines have proved to be invaluable for the creation of marker lines and for the isolation of regulatory sequences and eventually the isolation of specifically expressed genes including the isolation of environmental and hormonal stress responsive regulatory sequences and genes.

The promoter trap approach is particularly relevant to identification and cloning of genes (and their regulatory sequences) expressed in tissues that are difficult to analyze, by traditional methods relying on RNA extraction, as demonstrated by the cloning of regulatory sequences driving reporter gene expression in nematode feeding structures or the identification of molecular markers for embryogenesis. A T-DNA tagged mutant showing intense GUS expression in young leaves and rapidly growing stem tissues was used to isolate an eukaryotic translational initiation factor eIF-4AI. Promoter-like sequences, known as cryptic promoters which are not associated with a detectable transcript have also been reveaied by promoter trapping. The popularity of the promoter trap method can be judged from the various collections developed worldwide as well as the variety of reporters and constructs in use.

A collection of 20,261 transgenic lines of Arabidopsis generated with the promoter trap vector pTluc, which carries a promoterless firefly luc (luciferase) reporter gene linked to the right T-DNA border is the latest report in line. By detection of bioluminescence in three week-old seedlings, 753 lines were identified showing constitutive, organ-specific, and stress-responsive luciferase expression patterns. Several lines showed sugar, salt, and abscisic acid (ABA)-inducible luciferase activity. Moreover the technique has now been adopted for the ongoing intensive functional genomics efforts in rice demonstrated by the trapping of two cold inducible genes from screening of T-DNA tagged rice lines for cold responsive GUS expression.

Table 2 provides a list of promoter elements cloned using promoter trap vectors. Jeong et al. have created a total of 13,450 T-DNA insertion lines in rice that could be used to screen for promoter activity as well as for creating gain of function mutants.

Table 2. Promoter elements cloned using promoter trap vectors

S.No.	Promoter	Specificity of expression	Source species
1	At EM	Embryo	Arabidopsis
2.	Cryptic	Seed coat specific	Nicotiana tabacum
3	HVT1	Tapetum and vascular tissue	Arabidopsis
4	Pyk20	Nematode feeding structure	Arabidopsis
5	tcup	Constitutive	Nicotiana tabacum
6	Cryptic	Guard Cell	Arabidopsis
7	Cryptic	Roots	Arabidopsis
8	Lj Cbp 1	Roots	Lotus Japonicus
9	eIF-4A1	Growing tissues,	

		young leaves	Arabidopsis
10	EXORDIUM (EXO)	Meristematic cells	Arabidopsis

In addition to the promoter less GUS reporter gene located next to the right border, the binary vector used in the study carries multimerised transcriptional enhancers from the CaMV35S promoter near the left border. Interestingly histochemical GUS assays revealed the GUS-staining frequency in these lines to be about twice as high as that from lines transformed with vectors, which lack the enhancer elements suggesting that enhancer sequence present in T-DNA improves GUS tagging efficiency.

Isolating and Cloning Sequences Flanking T-DNA

Once co-segregation of T-DNA along with the phenotype has been established for a given mutant, isolation of the disrupted gene can be achieved by a number of strategies. Sequences flanking the insertion can be easily identified from single or low copy number lines using inverse PCR (IPCR) and by thermal asymmetric interlaced (TAIL) PCR (Figure 5).

The latter consists of three consecutive rounds of PCR amplification, performed with a set of three nested insertion-specific primers and a small, arbitrary primer, which anneals nearby in the insertion flanking sequence. Recently, Antal et al. have reported a simplified version of TAIL PCR called, single oligonucleotide nested (SON)-PCR, which involves only two rounds of PCR with two or three nested sequence primers.

Another strategy employed called, plasmid rescue makes use of insertion within T-DNA of an 'ori' sequence a replication of origin required for replication of plasmids in E. coli, along with a bacterial selection marker. Genomic DNA from mutant plant is digested with an appropriate enzyme and religated at high dilution to favour intramolecular reactions. The ligation products are introduced into E. coli and clones are selected on appropriate antibiotic selection. Because of the presence of the proper origin of replication and the presence of bacterial selectable markers, the plasmid is able to replicate in a proper E. coli host.

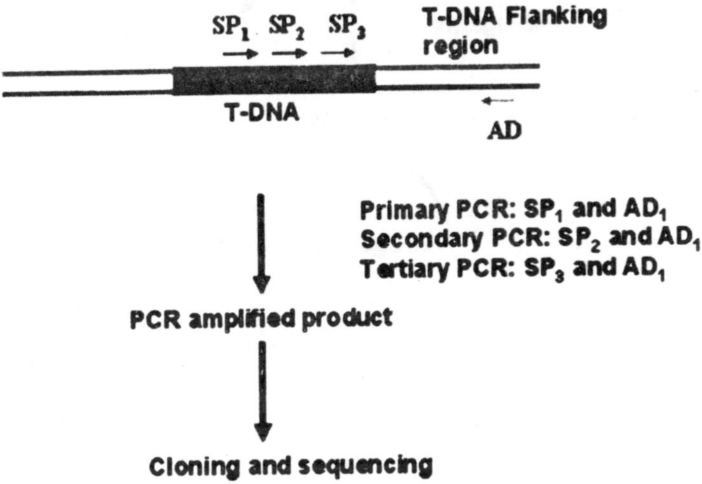

Figure 5. Schematic representation of TAIL PCR

The plasmid isolated from such bacteria thus contains the linked plant DNA as a passenger (Figure 6). A similar rationale applies to inverse polymerase chain reaction. In this procedure, plant DNA is digested with a restriction enzyme that cuts within the T-DNA and subjected to self-ligation. The circularized ligation products are subjected to PCR using appropriate primers from the T-DNA such that the flanking plant sequences, which are now part of the circularized product, are amplified (Figure 7).

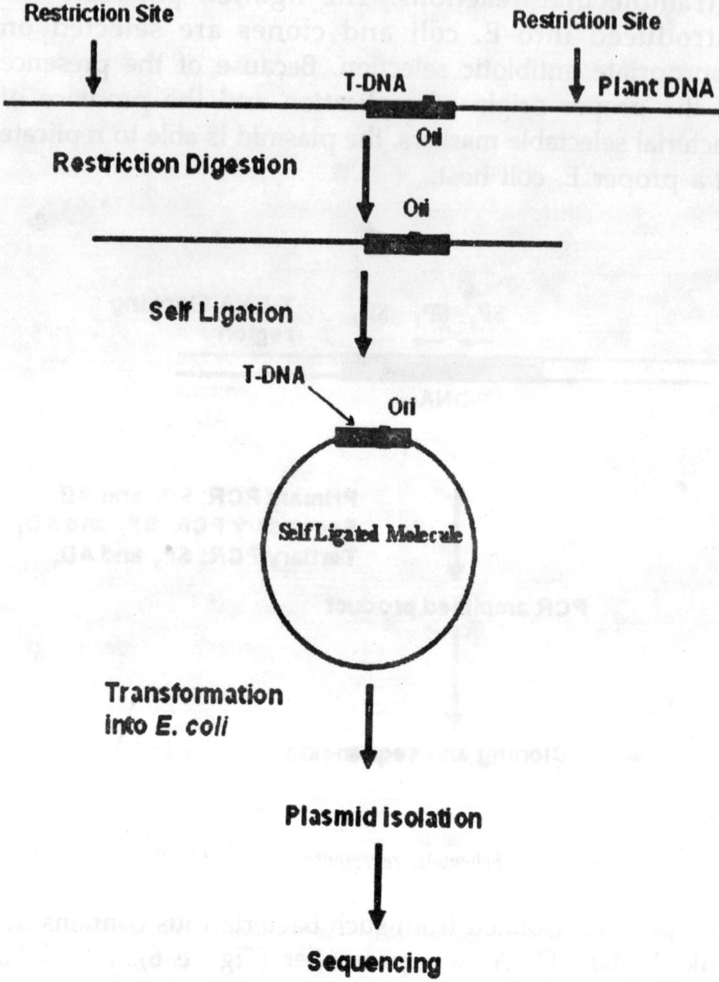

Figure 6. Schematic representation of plasmid rescue

Adaptor PCR or anchored PCR ligates adaptor molecules to the ends of the digested DNA fragments carrying part of the insert. A primer specific to the adaptor region and nested primers specific to T-DNA are used to amplify the flanking region (Figure 8). This method has been efficiently used in the German plant genome project GABI and flanking sequence database has been created using this data.

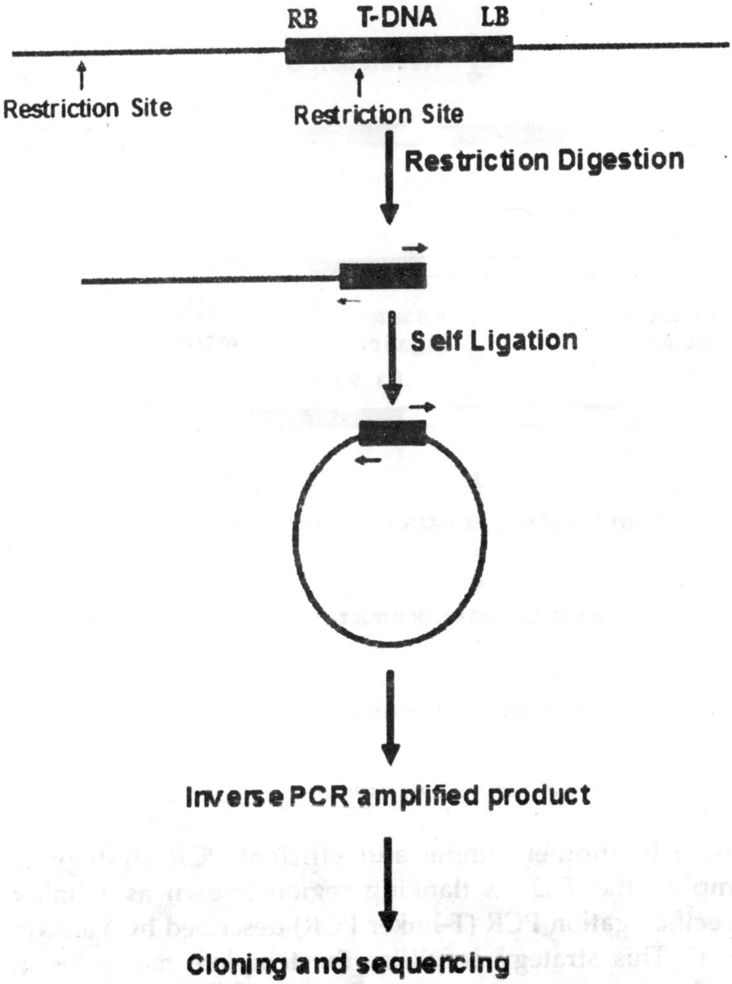

Figure 7. Schematic representation of inverse PCR

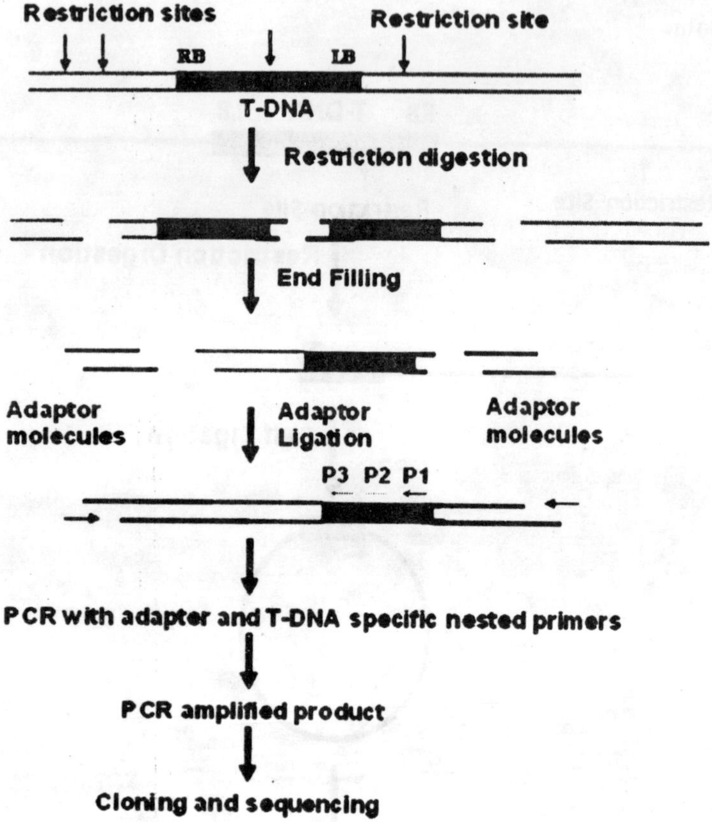

Figure 8. Schematic representation of adaptor PCR.

There is another simple and efficient PCR strategy to amplify the T-DNA flanking region known as T-linker specific ligation PCR (T-linker PCR) described by Yuanxin et al. This strategy amplifies the template molecules in three steps. First genomic DNA is digested with 3′ overhang enzymes. Secondly, primed with by a specific primer, a strand of target molecule is replicated by Taq

DNA polymerase and a single A tail is generated on the 3′ unknown end of the target molecule, and then a 3′ overhang-T linker (named T-linker) is specifically ligated on to the target. Thirdly, the target is amplified by two rounds of nested PCR with specific primers and T-linker primers. It differs from other methods in specific TA ligation instead of arbitrary ligation or random annealing.

Finally one can simply construct a genomic library of the mutant and isolate clones containing T-DNA sequences. Some of these clones will contain plant DNA as well. In any case unequivocal proof about the function of gene is obtained by performing molecular complementation, i.e., by demonstrating that the introduction of a wild type allele would eventually restore the phenotype.

Problems in T-DNA Lines

A variety of problems are encountered while analyzing the T-DNA tagged lines, in particular while attempts are made to clone flanking sequences from a T-DNA tagged mutant. Difficulty can arise because of multiple insertions, complex arrangement of T-DNA, insertion of vector backbone sequences, chromosomal duplication and rearrangements or a combination of these.

Multiple insertions and complex T-DNA loci

Complex T-DNA inserts composed of two or more T-DNA repeats are frequently found in transgenic lines. Methods like the TAIL PCR allow for the direct identification of insertions sequences co-segregating with the mutant phenotype even in the presence of more than one insertion. The flanking sequences representing the part of the gene of interest can be readily used to identify the corresponding gene from the Arabidopsis genome database.

Transfer of vector backbone

Although T-DNA tagging and isolation of flanking sequences may appear a rather simple and straight forward method to clone genes and promoters, but the inconsistency in T-DNA integration can complicate efforts to clone and identify adjacent plant DNA. Studies have indicated that Agrobacterium T-DNA can be present in the genomes of transformed host plants as single units or in multiple tandem arrays. Some of them have also shown the existence of truncated T-DNA regions.

T-DNA regions beyond the border repeat were also found to be stably integrated into plant genomes at high frequencies. Read through of border sequences during T strand formation in the bacterium has been reported. Even DNA sequences of the vector, from far beyond the defined T-DNA region delimited by the border sequences have been detected. Therefore, flanking sequence amplification from T-DNA tagged mutants has the potential of amplifying considerably larger tags of vector backbone sequences than expected.

This problem can be solved by designing primers outside the T-DNA borders and using special Taq polymerases capable of performing long PCR amplification. Long PCR techniques has become routine in most of the large scale T-DNA tagging projects owing to the relatively high prevalence of vector backbone sequences in the mutant population. These sequences transfer to the plant either independently of (unlinked to) the T-DNA or linked to the T-DNA across either the left or right T-DNA border.

75% of transgenic tobacco plants generated using Agrobacterium mediated T-DNA transfer contain binary vector backbone sequences integrated into the plant genome. Therefore, integration of backbone sequences besides the T-DNA sequences into the genomes of plant may also be responsible for causing mutations in T-DNA tagged populations. A number of groups have reported that in a large percentage of T-DNA tagged Arabidopsis plants, the T-DNA did not co segregate with the mutant phenotype.

Chromosomal duplication/rearrangement

Internal chromosomal duplication/rearrangements have been identified in the course of both reverse and forward-genetics approaches. An embryo-defective88 mutation was characterized with simple T-DNA inserts that exhibited normal Mendelian segregation and detected through a combination of detailed molecular analysis and mapping of the kanamycin-resistance phenotype imparted by the T-DNA. Although emb88is located on chromosome 1, molecular analysis of genomic DNA directly adjacent to the T-DNA left border revealed sequence from chromosome 5.

Mapping of the T-DNA confirmed that a >40 kb region of chromosome 5 had inserted with the T-DNA into the emb88 locus on chromosome 1. A similar scenario was observed in a reverse-genetic study of the LR-RPK gene. A prospective insertion mutation was identified by PCR screening of T-DNA lines. Whereas wild-type LR-RPK is encoded on lower chromosome 4, mapping of the T-DNA localized the mutant allele to chromosome 5. In both these cases, sequence of T-DNA flanking regions did not provide an accurate picture of DNA disruption.

Therefore, T-DNA insertion lines- even those that exhibit straightforward genetic behaviour with regard to T-DNA and developmental phenotypes may contain an unexpectedly high frequency of duplication/transpositions. Such rearrangements can interfere with reverse-genetic analyses and provide misleading information on the molecular basis of mutant phenotypes. This would necessitate, genetic mapping of T-DNA insertion sites, in addition to standard genetic and molecular characterization of mutant alleles for meaningful interpretation of phenotypes or sequence information from T-DNA mutant.

Insertional mutagenesis in Angiosperms

The progress of genomic analysis of plants has made huge leaps in recent years and a large body of information on genomic sequences, sequences of cDNA's and physical maps are now available for several plants including rice, maize, tomato and Brassica. Databases of many if not all, gene sequences of model plants are available and bioinformatics is enabling the prediction of gene function.

On the basis of comparisons among genomes, it is still necessary to validate these predictions. So various methods used for model Arabidopsis genome have been optimized and suitably streamlined for the systematic identification of the biological functions of genes in other more economically important crop species. Rice genome has been extensively covered because of its small size among cereal genomes, nearly complete genome sequence and economic importance, as a major crop. Maize Ac/Ds transposon system has been proved to be efficient for gene tagging and functional analysis in this crop.

Kolesnik et al. has reported a collection of unlinked single copy Ds transformants. Analysis of 2057 Ds flanking sequences has revealed preferential transposition of Ds into regions of genome rich in expressed sequences. This coupled with a high germinal and independent transposition frequency establishes the efficiency of this system for large-scale mutagenesis in rice. T-DNA tagging has been used successfully for gene discovery in rice. An initial database has also been constructed using T-DNA flanking sequences. T-DNA tags have also been observed to insert preferentially into gene rich regions.

In tomato new tools for functional analysis based on insertional mutagenesis, with Ac/Ds system in the background of the miniature cultivar Micro-Tom has been established. Micro-Tom is well suited for large scale mutagenesis in tomato owing to its small size, rapid life cycle, easy transformability, and efficient activity of the Ac/Ds elements. In addition, promoter trapping using firefly luciferase reporter gene and enhancer trapping using GUS was also developed. A high luciferase activity in the flowers, fruits and seedlings (69%) suggested preferential insertion of Ds elements into genes.

Data was also supported by flanking sequence analysis. This well initiated high –throughput functional analysis of genes and promoter isolation in tomato was complemented by a T-DNA insertional mutagenesis programme using activation tagging to identify genes that regulate metabolic pathways, which produced >10,000 independent activation tagged transgenic tomato lines and identified large number of leaf and fruit colour mutants. Analysis of an activation tagged insertion line with intense purple pigmentation led to cloning of ANT1 gene, a transcriptional regulator of anthocyanin biosynthesis, modification and transport.

T-DNA tagging strategy for isolation of new promoters has been tried in Brassica napus employing hypocotyl transformation with a promoterless gus: nptII tagging construct. This led to the isolation of new constitutive promoters suitable to drive expression of transgenes in crops as well callus specific promoters that can be used to drive selectable marker genes enabling selection in early phase of transformation procedure and eliminating marker gene products from mature transgenic. Considering the high efficiency gained by the reverse genetic programs it appears likely that all of the 20,000 to 25,000 genes that make up the basic angiosperm genome will be assigned function on the basis of experimental evidence by the turn of the decade. Understanding the molecular basis of developmental and metabolic diversity among angiosperms is the key to effective and rational improvements in the productivity and utility of crop species.

7

Transgene Expression

'Gene silencing' refers to the complete or partial inactivation of transgenes and homologous endogenous genes and is a frequently reported phenomenon in transgenic plants. Transcriptional silencing, characterized by a lack of transcript initiation and DNA methylation at promoter sequences, is very stable during mitotic and meiotic transmission. These epigenetic changes in gene expression are a risk for the successful application of gene transfer techniques in agriculture. In addition, they may be also relevant for differentiation and evolution in plants and represent an interesting mechanism of gene regulation.

There was a time not too long ago when most medicinal compounds came from plants: the potent heart stimulant digitalis from foxglove, for example, and opium from poppy plant. But beginning about 50 years ago, chemistry took over from botany, with most new drugs being artificially produced in pharmaceutical labs. Nowadays, one of the most promising methods of producing proteins and other medicinal substances, such as antibodies and vaccines, is the use of transgenic plants.

A transgenic plant contains a gene or genes that has been artificially inserted. The inserted gene sequence, known as the transgene, may come from an unrelated plant, or from a completely different species. One of the purposes of inserting a combination of genes in a plant is to make it as useful and productive as possible. This process provides advantages such as higher yield, improved quality, pest or disease resistance, and tolerance to heat, cold and drought. However, transgenic plants can also be produced in such a way that they express foreign proteins with industrial or pharmaceutical value. Transgenic plants represent an economical alternative to fermentation-based production systems. Plant-made vaccines or antibodies (plantibodies) are especially striking, as plants are free of human diseases, thus reducing screening costs for viruses and bacterial toxins.

The first transgenic plants were reported in 1983. Since then, many recombinant proteins have been expressed in several important agronomic species of plants including tobacco, corn, tomato, potato, banana, alfalfa and canola. The choice of plant system was initially driven by convenience and the need to evaluate genetic constructs quickly. For this reason, tobacco plants were generally used, however, potatoes and bananas were considered for the purpose of acquiring a delivery system of vaccines for humans.

Foreign gene expression in plants can be accomplished either by stable integration of foreign DNA, which results in transformation of the nuclear genome, or by transient expression using modified plant viruses. Stable genomic integration is accomplished by introducing foreign DNA in the plant either by Agro bacterium T-DNA vectors or by direct means. The integration method has the advantage of permitting large scale cloning, maintenance of selected high-expressing genes and the ability to sexually cross transgenes to obtain multiple proteins expressed in the same plant. On the other hand, transient expression using viral vectors is harder to initiate, because the viral vector must be inoculated into individual host plants, but gives a greater yield of protein.

Advantages and Disadvantages

Recombinant protein production using transgenic plants as bioreactors is likely to be more economical than alternative systems, especially for large-scale needs. Factors in favor of plant systems as sources of animal derived proteins, compared with other conventional methods, include:

— The potential for large-scale, low-cost biomass production using agriculture.

— Low risk of product contamination by mammalian viruses, blood-borne pathogens, oncogenes and bacterial toxins.

— The capacity of plant cells to correctly fold and assemble, not only antibody fragments and single chain peptides, but also full-length multimeric proteins.

— Low downstream processing requirements for proteins administered orally.

— Elimination of the purification requirement when the plant containing the recombinant proteins is edible, such as potatoes.

— The ability to introduce new or multiple transgenes by sexual crossing of plants.

— The avoidance of ethical problems associated with transgenic animals.

— Formulated in seeds, plant-made enzymes have been found to be an extremely convenient method for reducing storage and shipping costs, for an indefinite amount of time, under ambient conditions.

— Production size is flexible and easily adjustable to the needs of changing markets.

— Plants are also capable of synthesizing and assembling virtually any kind of antibody molecule, ranging from the smallest antigen-binding domains and fragments, to full length, and even multimeric antibodies.

— There are, however, potential issues of concern for plant protein production:

— Allergic reactions to plant protein glycans and other plant antigens.

— Plant and product contamination by mycotoxins, pesticides, herbicides and endogenous metabolites.

— Regulatory uncertainty, particularly for proteins requiring approval for human drug use.

Applications

The production of foreign proteins in plants has become an attractive alternative to conventional production systems for pharmaceutical polypeptides. Potential proteins produced by this method include cytokines, hormones, enzymes, epidermal growth factors, int rferons, human protein C, and pharmaceutical foodstuff considered for oral immunization.

The capacity of plants to produce different classes of proteins with pharmaceutical value and the need for new technology for the production and delivery of inexpensive vaccines has led to the use of transgenic plants. In addition, large amounts of antibodies can be produced at relatively low cost, using agriculture instead of sophisticated and expensive cell culture-based expression systems. Oral vaccines, as opposed to parenteral vaccines, offer the hope of more convenient immunization strategies and more practical means for implementing universal vaccination programs throughout the world.

The pathogens responsible for the greatest burden of human disease make initial contact with the human host at mucosal sites in the respiratory tract, gastrointestinal tract or genital tract. Therefore, stimulation of immune responses of suitable strength and quality to protect against illness is particularly desirable at these sites.

The best way of achieving mucosal immunization is to apply the vaccine directly to the mucosal surface, inducing systemic and cellular immune responses as well as local immune responses. Mucosal vaccines have been sought because of the potential for stronger immune responses directed at the initial site of interaction between the pathogen and host. Many infectious agents colonize or invade epithelial membranes. These include bacteria and viruses that are transmitted in contaminated food, water or by sexual contact.

Transgenic plants that express antigens in their edible tissue might be used as an inexpensive oral-vaccine production and delivery system. Therefore, immunization might be possible through consumption of an "edible vaccine" to provide passive immunization. Also, it may be possible to use genetically engineered plants and plant viruses to produce vaccines against several human diseases from tooth decay to life-threatening infections such as diphtheria, cholera and AIDS.

Several issues will still have to be resolved before this intriguing idea can become a reality. Some of the proteins used in experiments may be extraordinarily potent inducers of immune responses. On the other hand, other immunizing proteins may not work as well when taken orally. In fact, they can have the opposite effect, because many proteins in a diet induce tolerance, making the immune system less able to mount a response against them. In addition, some other compounds in plants may compromise the ability of the vaccine protein to induce immunity. The food containing it must be palatable and some foods need to be heated before ingested, which could possibly cause the vaccine protein to denature, reducing or eliminating its ability to elicit immunity.

Oral vaccines, whether living or nonliving, must be protected during passage through the hostile environment of the stomach and intestine to the sites where immune stimulation occurs. In recent years, a variety of delivery systems have been developed for presenting nonliving antigens to mucosal surfaces, which will allow these antigens to persist and survive in the hostile gastric and enteric environments. These include polylactide/ polyglycolide, microspheres, liposomes, proteosomes, cochleates, virus-like particles, and immune-stimulating complexes.

One of the most promising methods, however, is the production of antigens in the plants themselves, which assemble into ordered structures such as virus-like particles. This gives the hope that they will be more resistant to digestion and more likely to reach the gut-associated lymphoid tissue.

Methods of Production

Conventional Method

Pharmaceutical and therapeutic antibodies synthesized in plants can be produced in a variety of ways. Conventional methods use stable transformation and transient expression to introduce new genes into a host cell. Once DNA from the transformant host cell is isolated and purified, it can be injected into the embryo of a maturing plant. The plant can then propagate in an open field allowing for large-scale production of antibodies. As mentioned before, however, purification of these proteins is generally long and tedious. Upon isolation of the antibody, several proteins, organic molecules, glycan and herbicides must also be isolated, leading to a complex purification process.

In Vitro Cell Tissue Cultures

Plant tissue cultures offer an economically favorable method for producing antibodies from plants. Using this approach, plant cells in differentiated or dedifferentiated states are grown in a nutrient medium in bioreactors under controlled conditions, with foreign proteins harvested from either the biomass or culture liquid or a combination of both.

This method of production of human antibodies is not suitable for the production of edible vaccines (simply because the antibody is produced in a cell culture and not in a fruit or vegetable) but offers many advantages to the conventional methods of extracting and purifying a protein from a live plant. First, plant tissue cultures offer larger amounts of proteins in shorter amounts of time. This is because the bioreactors provide a much more controlled and reproducible environment than an open field.

The advantages of this method allow desired antibodies to be produced, purified, and transferred to the consumer in a minimal amount of time. Also, purification of the proteins becomes easier. In vitro cell cultures contain fewer biological proteins or molecules (along with herbicides and pesticides) than open field plants or bacterial/yeast cell cultures, which may contaminate the product. Furthermore, plant cells and organs can propagate indefinitely in tissue cultures.

Therefore, sexual reproduction is not needed to ensure the lifespan of the species. Without sexual reproduction, transgene stability is increased because of the absence of crossing over, segregation and recombination involved in sexual reproduction.

Inducible promoters may offer a solution to the problems associated with open field production of plantibodies. They would allow for better efficiency in the production of plantibodies. They would also provide the plant with a more stable mode of translation, leading to transgene stability.

Breeding and Sexual Crossing

In 1989, Hiatt was the first to use sexual crossing as a way to produce a functional protein in plants. Transformation was used to introduce kappa-chains of either light or heavy regions into tobacco plants. The same was done with gamma-chains of either light or heavy regions. Upon crossing one plant with kappa-chains and another plant with gamma-chains, an antibody was produced that expressed both chains.

Transgenic seeds

Using green plant tissue as a bioreactor is a great method of producing antibodies. Plants cannot store antibodies for an extended period of time. This is because certain proteases degrade the protein piece by piece. Ulrike Fieldler and Udo Conrad showed that seeds of transgenic tobacco could be used for high-level production and long-term storage of antibodies. In their experiments, functionally active single chain Fv (scFv) accumulated up to 0.67% of the total soluble seed protein.

In addition, storage of ripe transgenic tobacco seeds for one year at room temperature showed no loss of scFv or its antigen-binding activity. Seeds contain a low level of proteases that allows proteins to be stored without degradation.

Targeting and Compartmentalizing

In some cases, antibodies can be tagged with a small peptide sequence. This peptide sequence allows plant cells to direct an antibody or protein to specific organelles and compartments after processing. Compartmentalizing to easily isolated organelles provides a less complicated purification procedure. Targeting also allows antibodies to be protected from proteases that exist in the cytoplasm of cells. Targeting, however, has to be specifically controlled. This involves proper cleavage of the targeting sequence. If incomplete processing occurs, the quality and amount of protein is lowered.

Methods of Suppression

Gene Silencing

Recent studies with transgenic production of antibodies have shown that the transgene involved can undergo inactivation; this process is termed gene silencing. Gene silencing has been observed to occur where there exist multiple copy integrations at one or more sites, different base-composition between rDNA and the integration site, detrimental effects of sequences adjacent to the rDNA integration site and over-expression effects.

There are certain ways to prevent the occurrence of inactivation: Screening/selection for plants with single copy rDNA; developing methods for single-copy integration; avoiding repetitive homologous sequences; flanking rDNA with scaffold attachment regions; selection/screening for stable rDNA expression; and developing site-specific recombination systems.

Oral Tolerance

When vaccines are taken orally, either with potato tubers or with bananas, the intestinal immune response to these food-antigens is called tolerance. Oral tolerance is an active immunologic response, which prevents the development of responses to the many antigens ingested by a host. Upon intake, the oral antigens are taken up by M cells. These M cells present antigens to specific suppressor T cells, which secrete cytokines that allow for the propagation of specific antibodies. Tolerance is dose-schedule-, and antigen-specific. Therefore, an antigen that is orally administered and is tolerant will show no response, because the expression of that antigen has been suppressed. When synthesizing a protein from a transgenic plant, oral tolerance must be taken into consideration.

Creating SIgA

Secretory immunoglobulin A (SIgA) is the most abundant form of immunoglobulin in mucosal secretions. It is composed of two monomeric IgA antibody units, a small J-chain and a secretory component. In an experiment, which four plants expressing each of the polypeptides were sexually crossed, it was found that a functional SIgA antibody could be produced. The functionality of the protein was studied by enzyme linked immunosorbent assay (ELISA). SA I/II was used as the antigen to be recognized on the transgenically produced SIgA antibodies. SA I/II is a cell surface protein of Streptococcus mutans, which can cause dental caries in humans.

CaroRxTM: An anti- S. mutans SigA CaroRxTM is a clinically advanced SIgA plantibody that protects humans from dental caries. In a preliminary study of CaroRxTM, anti-S. mutans antibodies were orally administered to 84 human subjects. Upon the application of SA I/II, monoclonal antibodies prevented the colonization of artificially and naturally implanted S. mutans. In addition, protection from recolonization (with just 3 weeks of application) lasted for two years.

Production of an Oral Cholera vaccine

From 1995 to 1998, extensive research was conducted in the usage of potatoes as an antigen delivery system to initiate an immune response. Of particular interest was the CT or cholera toxin antigen. CT is composed of an A subunit and B pentamer subunit. CTA is an enzymatically active protein, which enters epithelial cells of the gut and causes water loss from these cells. CTB is enzymatically inactive and binds to Gm-gangliosides of epithelial cells, which allows for the intake of CTA.

In one experiment, Arakawa, Chong and Langridge used mice to study the effects of a CT oral vaccine. The mice were fed transgenic potato tissue containing thirty micrograms or CTB/gram. They were fed four times, on days zero, six, seventeen, and twenty-four, and received a booster feeding on day sixty-five. Ten mice were fed one gram of transgenic potato expressing CTB while eight mice were fed three grams of transgenic potato expressing CTB. In addition, five mice acted as a negative control and were fed one gram of untransformed potato while eight mice were used as a positive control and fed thirty micrograms of bacterial CTB in sodium bicarbonate buffer.

It was observed that during the 70-day experiment, mice immunized with thirty micrograms of bacterial CTB showed 55% protection. In addition, there was 42% protection for mice immunized with one gram of transgenic potato, and 62% protection for mice immunized with three grams of transgenic potato. From these experiments, it could be concluded that oral delivery of antibodies was an effective way to treat certain diseases, at least in animals. Similar experiments were done with LT-B or heat labile enterotoxin, which showed similar results.

Transgene Expression and Silencing in Arabidopsis

A molecular genetic approach in *Arabidopsis* is taken to elucidate the relationship between genome structure and gene expression. Introduced genes often show variable expression - in some cases, they are even completely silenced. Neither the locus of integration nor the copy number of the introduced piece of DNA can be efficiently controlled beforehand if transgenic plants are being generated.

Therefore, we wish to understand how copy number and/or position of a transgene in the genome influence its expression. Transferred DNA (T-DNA) lines harbouring reporter genes of distinct types, defined numbers and arrangements in many different genomic positions under the control of a promoter conferring high constitutive expression.

Below a certain number of identical transgenes, we observe that each gene copy contributes equally to gene expression. This expression is high, stable in all generations analysed, and comparable between independent lines carrying the same transgene copy number. Thus, neither position effects nor repeat arrangements of transgenes per se account for the large variability of transgene expression and/or cause silencing seen among transformants. Conversely, RNA silencing is triggered if the transcript level of a transgene surpasses a gene-specific threshold.

These results suggest a model that is of immediate relevance to all over-expression studies in *Arabidopsis* and to studies that exploit reporter gene fusions. The expression of highly similar genes is additive unless transgene transcript levels exceed a gene-specific threshold and consequently trigger post-transcriptional gene silencing (PTGS). Positive gene dosage effects and, in particular, the stochastic onset of silencing effectively account for the large variability of transgene expression seen among transformants.

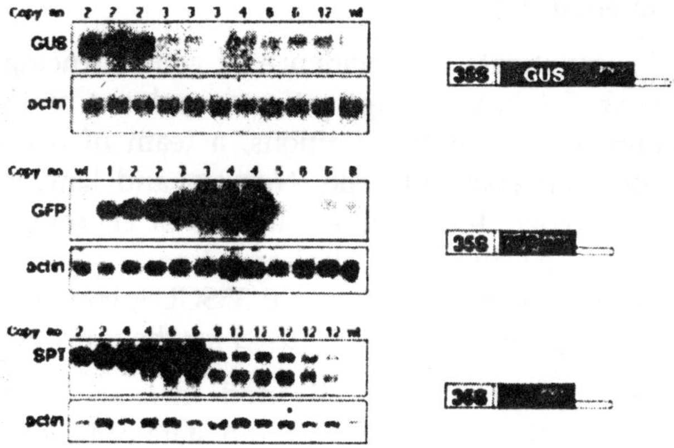

Figure 1. Post-transcriptional gene silencing

Environmental Influences on Gene Silencing

Gene silencing may be important for inactivating deleterious foreign genes and for normal growth and development of plants. However, plant genetic engineers do not welcome the phenomenon, as it interferes with the expression of newly inserted genes. Many factors affect the expression of gene silencing. Reports indicate that it can occur at either the transcriptional or posttranscriptional level.

Silencing that occurs at the transcriptional level may be caused by direct physical association or pairing of alleles. Also, the ploidy of the plant may sometimes influence gene expression, since reduced transgene expression is observed in triploid, as compared to diploid, Arabidopsis hybrids. Posttranscriptional transgene silencing is affected by both plant developmental stage and environmental factors. The experimental data accumulated to date do not fully explain the mechanism of gene silencing, nor do the suggested models accommodate all experimental data.

To investigate the incidence of gene silencing using different T-DNA configurations under a variety of environmental growth conditions, a team of researchers from the University of Oslo, Norway and University of Skovde, Sweden developed a collection of 111 independent transgenic Arabidopsis thaliana lines. Two vectors, pPCV002 35SGUS and pKOH110 35SGUS, containing gus and nptII reporter genes, were used for the transformation.

The 111 lines were divided into six categories according to the number of loci and state of hemizygosity or homozygosity using segregation ratios for kanamycin resistant and kanamycin sensitive phenotypes in T2 seeds. Segregation data were used to determine the extent that nptII-silencing in the different categories was different from the lines screened.

For each transgenic line, eleven four-week-old seedlings were selected and kept at normal growth conditions for three days. Then stress treatment, consisting of 30°C day and night temperatures, was given to seven plants, of which four were also sprayed with insecticide. The other four plants were given a 30°C stress treatment during the day and 4°C during the night. After two weeks, the plants were transferred to original growth conditions (22°C). NptII-silencing was scored by planting dormancy-broken surface-sterilized T3 seeds. Determinations of T-DNA copy number and DNA methylation were made using Southern hybridization.

Sixty-seven and 44 transformants were independently generated for pPCV002 and pKOH110 vectors, respectively. NptII-silencing was scored for each line after germination on medium containing kanamycin. Three silencing phenotypes, I - III, were identified. White or light green cotyledonous plants were designated as Type I; those with white, spotted, and deformed leaves were classified Type II; and Type III had large, green, spotted leaves.

Type II and III phenotypes are never found in wild-type seedlings. It was assumed that the white parts of the seedlings were displayed where nptII was silenced, inhibiting normal seed development. Those seedlings displaying nptII-silencing from the stress-treated groups were identified and the overall silencing frequencies of the stress and control groups were calculated for each line. The extra stress treatments, 4°C and insecticide, did not result in additional effects.

NptII-silencing of T3 plants was found in 56% of the 111 transgenic lines. In the majority of lines having a low to medium overall frequency of nptII-silencing (less than 10 - 50%), the frequency was higher in the stressed group than the control group; however, this was not seen for any lines displaying more than 50% silencing frequency. Differences in silencing frequency were not correlated to the T-DNA vector source.

Interestingly, the silencing frequency differed between siblings in all but two lines. In the majority of lines, the frequency of nptII-silencing was significantly different between the stressed group and the control. In several cases, the frequency of silencing in progeny of stress treated plants was higher than for the control group, while in other lines this trend was reversed.

There was a correlation between homozygosity and silencing in progeny of hemi- and homozygous siblings of the T3 generation and found that silencing could be attributed to homozygosity in only one line. Likewise, a correlation between kanamycin resistance and methylation of the Sac II site in the pnos promoter region, as has been reported in other studies, was found in multi-copy, but not single copy, lines. Additionally, the team found that the frequency of silencing was not correlated with rearranged transgenes.

Stress-induced susceptibility of plants to gene silencing is exhibited at three levels: (i) stress can change the proportion of sibling plants producing silenced progeny; (ii) individual plants may influence the number of seeds in which the transgene is silenced due to a change in the number of cells with silenced transgene(s); and (iii) silencing phenotypes can be altered. All transgene copies in all cells of Type I seedlings are likely to be silenced. However, in seedlings of Type II and III phenotypes, which contain seedlings that have developed beyond the cotyledon stage, transgenes are generally only partially silenced.

Meza et al. report that environmental stress can have either a positive or negative influence on the frequency of silencing. One transgenic line, which exhibited an increase in silencing after stress treatment, displayed both Type I and II phenotypes, while the control seedlings displayed the II and III types. Conversely, in another line in which stress treatment produced a reduction in nptII-silencing, the silenced progeny from the stressed group showed the Type III phenotype, and the control group displayed Types I and II.

Studies of the sequences flanking the T-DNA in one-locus lines, some of which contain a single copy, may help generate transgenic lines with stable expression patterns. In addition to questions concerning the mechanisms underlying gene silencing, there is interest in isolating genes whose products modify the timing of silencing. There is also a need to clarify the involvement of methylation in silencing and to minimize this effect by targeting transgenes into compatible isochores that include flanking scaffold attachment regions in the constructs.

Molecular Analysis of Rice Genes

The ultimate goal of rice biotechnology research is to produce superior transgenic plants that give stable expression of an introduced gene (transgene). Over the last several years, gene silencing in transgenic plants has emerged as a serious problem. In gene silencing, either the transgene is not expressed at all, or it is expressed in plants during earlier stages of growth but not in later stages, or it is expressed in the first generation (R_0) but not in the next generation (R_1). In general, gene silencing is often associated with integration of multiple copies of the introduced gene. The higher the copy number, the more likely the occurrence of gene silencing.

Gene silencing is also known as inactivation of homologous transgene expression. Several different mechanisms are thought to be involved in gene silencing, which include cis inactivation (frequently due to methylation of DNA), trans inactivation, or co suppression (sense suppression) of a transgene or a homologous endogenous gene. Gene silencing may also be related to chromosomal insertion sites of the transgene (position effect), or due to degradation of RNA-RNA duplexes.

One way to minimize the problem of gene silencing and to increase transgene expression is to flank the transgene with a DNA sequence known as a scaffold attachment region (SAR) or matrix attachment region (MAR). It has been proposed that when MAR sequences are present on both sides of a transgene, they can form a separate transcriptional active loop domain by binding to the nuclear matrix. Even though results from different groups, using different MAR sequences, are not always consistent, it seems that including flanking MAR sequences in plasmid constructs might be useful. In addition, eliminating repeated sequences from plasmid constructs may alleviate problems related to DNA-DNA pairing and de novo methylation. To minimize RNA turnover induced by excess RNA accumulation, moderate transcription rates (driven by a moderately strong promoter) may be preferable to very high rates (driven by a very strong promoter).

Gene silencing in transgenic rice was reported recently at the *General Meetings of the International Program on Rice Biotechnology,* September 15-19, 1997, Malacca, Malaysia, which was organized by the Rockefeller Foundation. Kumpatla *et al.* reported that direct DNA delivery procedures resulted in plants that contained multiple copies of the transgene, some of which were rearranged. By following *bar* gene expression, it was found that inactivation of the transgene was due to methylation. The silenced state was inherited, and *bar* gene expression was restored by germinating R1 seeds in the presence of 5-azacytidine.

Kloti *et al.* found that gene silencing occurred in the T_0 generation or later. In some cases, homozygosity enhanced the silent phenotype. With the RTBV promoter, cell-specific silencing was observed, which leads to lack of expression in the vascular tissue, but not in the epidermis. In these plants, DNA methylation of the promoter fragments was detected, and silencing was reversed by sexual crossing with nontransgenic plants or by treatment with 5 azacytidine.

Thara *et al.* observed that transgene silencing occurred in some T_3 and T_4 generation progeny of homozygous rice plants expressing an introduced chitinase gene. Viegas *et al.* found that among five R_0 plants expressing the reverse tran-scriptase gene (RT) of RTBV, one of them no longer produced RT mRNA in the R_1 generation. Since several copies of the same transgene are present in this R_1 line, the results suggest the possibility of gene co-suppression.

From the above examples, it appears that gene silencing is a problem that can arise from attempting to produce transgenic rice plants that stably express the transgene for many generations. It is likely that gene silencing can be largely prevented if only one or two intact copies of the transgene are integrated into the genome without the presence of rearranged copies. One solution to this problem is to use MAR sequences to flank the gene of interest to be introduced into rice cells by transformation.

8

Seed Germination and Plant Tissue Culture

A seed certainly looks dead. It does not seem to move, to grow, nor do anything. In fact, even with biochemical tests for the metabolic processes we associate with life (respiration, etc.) the rate of these processes is so slow that it would be difficult to determine whether there really was anything alive in a seed.

Indeed if a seed is not allowed to germinate (sprout) within some certain length of time, the embryo inside will die. Each species of seed has a certain length of viability. Some maple species have seeds that need to sprout within two weeks of being dispersed, or they die. Some seeds of Lotus plants are known to be up to 2000 years old and still can be germinated.

Assuming the seed is still viable, the embryo inside the seed coat needs something to get its metabolism actived to start the embryo growing. The process of getting a seed to germinate can be simple or complicated, and this our present subject.

Common vegetable garden seeds generally lack any kind of dormancy. The seeds are ready to sprout. All they need is some moisture to get their biochemistry activated, and temperature warm enough to allow the chemistry of life to proceed. Seeds taken from the wild, however, are frequently endowed with deeper forms of dormancy.

Dormant Embryos

There are several mechanisms that permit seeds to be truly dormant.

Thick Seed Coat

Many kinds of seeds have very thick seed coats. These obviously keep water out of the seed, so the embryo cannot get the water needed to activate its metabolism and start growing. The lotus seeds are an example of this. An outstanding example from the northern temperate zone is the Kentucky coffee tree (*Gymnocladus dioica*). The seed coat is perhaps two millimeters thick! You can throw them as hard as you can against a concrete sidewalk and they just bounce! How could such a seed actually sprout?

The Kentucky coffee tree holds its seed pods in the the top of the tree all winter. The inside of the pod is fleshy (lots of water). The pods are very dark in color. You might want to recall what happens if you fill the ice cube trays in your freezer too full with water, or you might recall what happens to a container filled with soda that is then frozen.

Other species might use some pounding along a river or drop seeds into seacoast surf to abrade the thick seed coat. Some of the sea beans do this. Other seeds might need an vertebrate or other animal to attack the seed coat (but give up trying to eat the seed) and thereby weaken the coat. The process of nicking the thick seed coat to initiate germination is called scarification.

A final, and very common, example of a way to scarify a seed coat is observed in strawberry and raspberry.

The thick seed coat is designed to be swallowed by the frugivore. The animal digests the fruit pulp, but the seed coat passes through the digestive system still protecting the viable embryo inside, but weakened enough to allow sprouting! The seed is deposited with a little organic fertilizer in the environment and can now sprout!

Thin Seed Coat

A thin seed coat is so thin that it is no barrier to water. Some other kind of dormancy mechanism is needed. Knowing that light can penetrate thin layers of plant tissue (leaves for example) should give you the idea that light might be a signal. That plants can absorb light and respond biochemically is a fact you know from your study of photosynthesis. All we need is a pigment molecule that can absorb light and cause a change in the behavior of the embryo.

The pigment is phytochrome. Like chlorophyll, it is made of a chromophore with tetrapyrole structure and is associated with proteins. This pigment is different from chlorophyll, however, in one critical way. It exists in two inter-convertible forms.

One form of phytochrome, named Pfr, is the form of the phytochome found in plant cells. This form of phytochrome is biologically very active and plays a role in all systems when a plant needs to know if the lights are "on" or "off." In lettuce (*Lactuca sativa*) seeds, Pfr causes the seeds to begin to germinate as we will soon see. Thus lettuce seeds germinate only when placed in white or red light. Buried in deep soil, they will not germinate. Given that lettuce has a small seed, you can figure out why evolution arrived at this solution.

The other form of phytochrome, named Pr. This form is biologically inactive or inhibits responses. Thus if lettuce seeds are placed in far-red light they do not germinate.

Large seeds have lots of storage material. If their seed coat is very thin, their evolution may have arrived at a completely different response. Think about pea seeds. They are large and have very thin seed coats.

Insufficient Development

If a seed's embryo is not completely developed, some additional maturation may be needed before the seed can sprout. This happens in seeds with little-to-no storage material invested in the seed. Examples include orchid seeds. They are the size of dust and have almost nothing but a very immature embryo on-board. Such a seed needs an association with fungi in the soil or other environments to feed the developing embryo until the embryo is mature enough to actually penetrate the seed coat. These seeds are also likely to have a very brief viability. The fungal association must be established rapidly or the embryo dies.

Inhibitors Present

Many plant species invest chemicals in the developing seeds, and these chemicals inhibit the development of the embryos. They keep the embryos dormant. Obviously the seed must have some way to eliminate these chemicals before they can sprout.

Abscisic Acid

Many temperate zone species that use inhibitors use abscisic acid. This chemical induces dormancy in the embryo. The chemical is produced in abundance in the late summer and early fall. The seeds in the fruits become dormant so, even if they are dispersed in autumn, they cannot sprout. During the winter enzymes in the seeds degrade the abscisic acid. By spring the abscisic acid is gone and the seed can sprout.

We can collect seeds of these species and get them to sprout early. The seeds are put in moist soil and refrigerated for about four weeks (a process often called stratification). This is sufficient time to degrade the abscisic acid. Then the planted seeds are placed in a warm greenhouse. The seeds assume winter is over, spring has come, and they begin to sprout. This process is called vernalization. If you think of "vernal" as meaning "spring" then you understand how we got this name!

Phenolic Compounds

Plants that live in deserts have a different problem. There is no cold, moist, winter to allow vernalization of abscisic acid. These plants instead use more potent toxins, phenolic compounds, to keep their seeds dormant until the proper season for germination. Phenolic compounds are freely water-soluble, the plant is living in a desert. Deserts typically have very long dry seasons and a short wet season accompanied by flash floods and so on. How do you think the phenolic compounds are lost? How would the mechanism ensure that seeds do not sprout in the dry season, but only after the seed could be sure it is in the wet season? The word leaching might give you a hint?

Dormancy and Germination

The seeds are small with a minute embryo surrounded by oily endosperm. Seed dormancy can be considerable, with delayed germination a common problem. B.R. Atwater classifies seed morphology as endospermic seeds with a basal rudimentary embryo. Potassium nitrate, gibberellin and pre-chill treatments generally promote seed germination.

The first step in the algorithm is to test seeds at constant temperatures of 11°C, 21°C and 31°C with light applied for 12h/d. If full germination is not achieved in any one regime and the results suggest a trend in the response of germination to constant temperatures then test at a more extreme or an intermediate constant temperature as appropriate.

For example, if germination at 11°C is greater than germination at 21°C or 31°C then test a further sample of seeds at a constant temperature of 6°C with light applied for 12h/d. A further example: if germination at 21°C and 31°C is roughly similar but greater than at 11°C then test a further sample of seeds at a constant temperature of 26°C with light applied for 12h/d.

If these constant temperature regimes do not promote full germination then the second step of the algorithm is to test seeds in alternating temperature regimes. If germination at constant temperatures of 21°C and below was greater than at constant temperatures above 21°C in step one then test a further sample of seeds in an alternating temperature regime of 23°/9°C (12h/12h) with light applied for 12h/d. If germination at constant temperatures of 31°C was greater than at all other constant temperatures in step one then test a further sample of seeds in an alternating temperature regime of 33°/19°C (12h/12h) with light applied for 12h/d. If germination at constant temperatures of 21° to 31°C was similar or if germination at a constant temperature of 26°C was greater than at all other constant temperatures in step one then test further samples of seeds in both the alternating temperature regimes described above.

If the second step of the algorithm does not result in full germination then the third step is to co-apply 7 x 10-4 M GA3 to the germination test substrate and test a further sample of seeds in the most successful temperature regime determined from a comparison of the results of steps one and two.

If the third step of the algorithm does not result in full germination then the fourth step is to pre-chill a further sample of seeds at 2° to 6°C for 8w and then test in the most successful regime determined from a comparison of the results of steps one to three. This may include co-application of GA3 if the results of step three show a significant increase in germination over the corresponding test in step one or two.

If full germination has not been promoted, the fifth step of the algorithm is to estimate viability using a tetrazolium test. If the result of the tetrazolium test indicates that the failure to achieve full germination is due to the presence of dead seeds and that one of the above regimes promoted the germination of all, or almost all, the viable seeds, then this regime is used for all subsequent germination tests. If, however, the result of the tetrazolium test indicates that dormancy has not been broken by the regimes applied so far in the algorithm, then experiment with modifications to the above regimes.

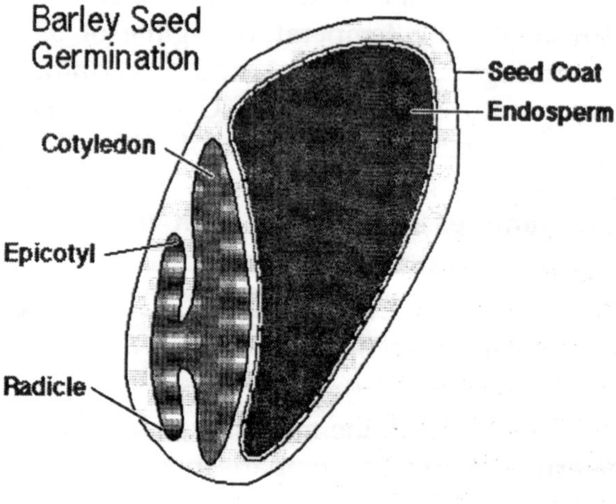

This is a diagram of the seed of barley (*Hordeum jubatum*)

As any "typical" seed it has three fundamental parts:

— a seed coat

— a storage area (in this case the endosperm) and

— a dormant embryo

The seed coat is really a fruit coat. In all grains, the seed coat (former ovule integument) is fused to the ovary wall (the true fruit wall). So in fact, the grain is technically a fruit (caryopsis) even though we often call it a seed.

Barley is used primarily in beer making. Brewers use barley as a source of sugar to make the alcohol for beer. Brewers knew that sprouting the barley seeds improves the sugar yields tremendously, but they wanted even more sugar yield. The brewers came to plant physiologists to find out if anything could be done.

Study of the seed showed that the seed/fruit coat is water-resistant and thereby reduces the rate of water uptake by the seed, and water uptake is essential for seed germination and the improved sugar yields.

The sugar content of a dry barley seed is actually quite low, but the endosperm holds a huge reserve of starch. Starch is a polymer of sugars. In a sense it is a long chain of sugar molecules linked together. This is probably the source of the sugar.

The barley embryo has three parts:

— the cotyledon (or seed leaf). Since there is only one, barley is a monocot.

— the epicotyl (becomes the shoot)

— the radicle (becomes the root)

Seed germination is said to have occurred when growth of the radicle bursts the seed coat and protrudes as a young root. The energy for seed germination probably comes from respiration of the sugar in the endosperm. However, the embryo and the starch are separated from each other. There must be some chemical communications that physiologists could manipulate to enhance sugar yield for the brewers.

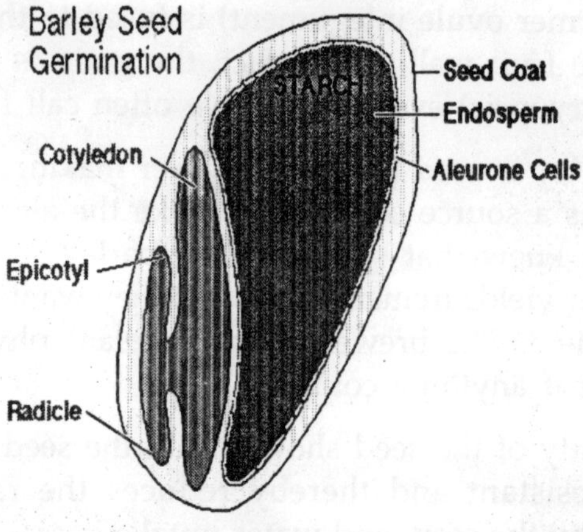

The endosperm of the seed has two parts. The bulk of the volume is a starch storage area. The covering layer is called the aleurone layer. The aleurone is made of cells that store protein in abundance.

You probably know about this in the case of rice. Rice grains are treated before we buy them for cooking. The fruit/seed coat is milled off and discarded as it has little food value. The embryo breaks out and is called the "germ." In wheat, the embryos are sold as wheat germ. The milling process leaves the rice endosperm surrounded by aleurone. This is sold as brown rice and has food value in starch and protein. It has an interesting texture after cooking. If the endosperm is polished before it is sold, the aleurone can be removed. Polishing results in white rice which has food value in starch only. Thus white rice, while considered much "finer" than brown rice, is actually a lower-quality food.

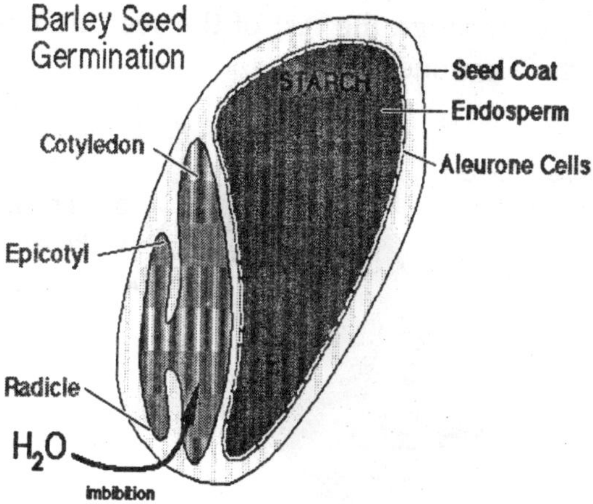

The first step in barley seed germination is imbibition. In this process, water penetrates the seed coat and begins to soften the hard, dry tissues inside. The water uptake causes the grain to swell up. The seed/fruit coat usually splits open allowing water to enter even faster. The water begins to activate the biochemistry of the dormant embryo.

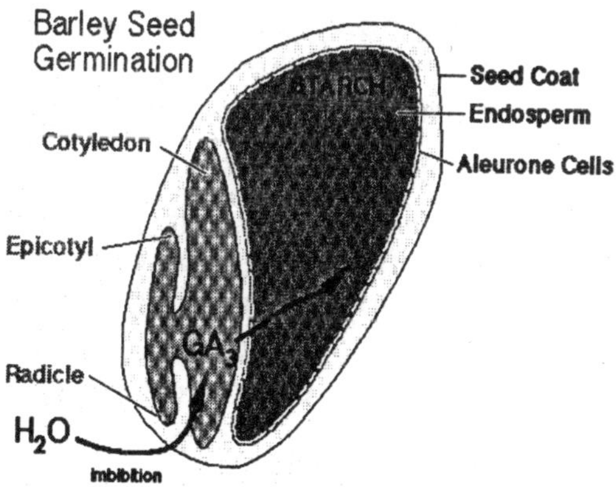

The water coming into the seed and embryo dissolves a chemical made inside the embryo. This chemical is called Gibberellic Acid (GA). It is a plant hormone, not too different from steroids. The dissolved GA is transported with the water through the rest of the seed tissues until it arrives at the aleurone layer.

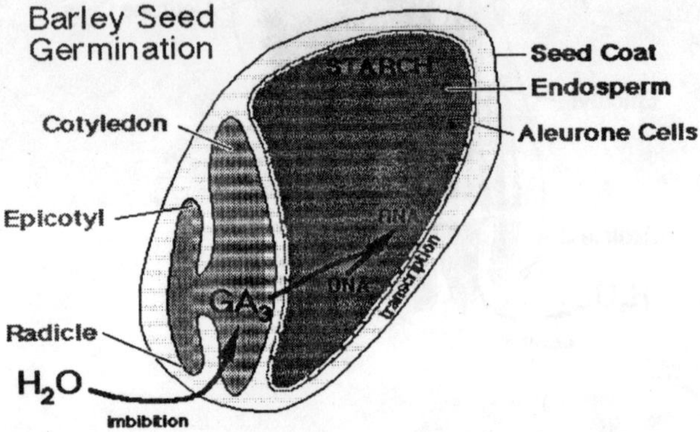

The Gibberelic Acid which is transported by the water arrives at the aleurone layer. The GA crosses into the cytoplasm of the aleurone cells and turns on certain genes in the nuclear DNA. DNA is, of course, the hereditary molecule and contains the instructions for making every protein needed for the survival of a barley plant. The precise mechanism of how GA turns on the DNA is unknown at present. It is clear however that the mode of action is to turn on just *certain* genes in the DNA.

The genes that are turned on are transcribed. The information archived in the DNA is precious, so the aleurone cells make a disposable RNA copy of the gene that is turned on. This disposable copy of the information, a kind of blueprint, is often called messenger RNA. The process of making this RNA copy is called transcription.

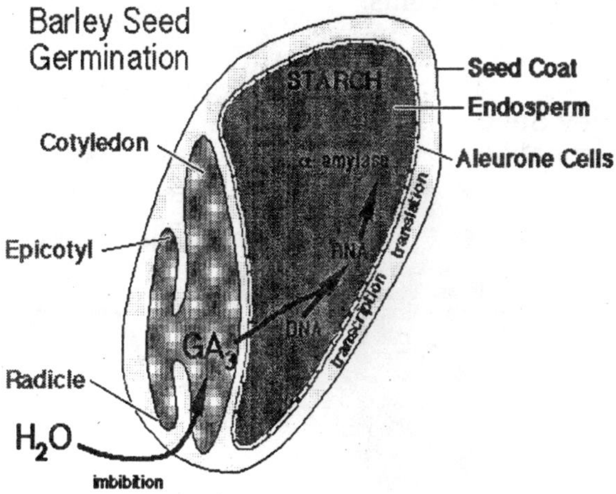

The RNA that was made in the transcription process is transported into the cytoplasm of the aleurone cells.

In the cytoplasm, the messenger RNA joins up with a ribosome to begin the process of making a protein. This process is often called protein synthesis or translation. In this process, the ribosome examines the information held in the sequence of bases in the RNA.

Transfer RNAs charged with particular amino acids are moved into the position specified by the instructions in the messenger RNA, and the amino acids are joined in a proper sequence by the ribosome. The sequence of amino acids determines the properties of the protein being assembled.

In this case, the critical protein made with the information held in the RNA is amylase. This protein turns out to be an enzyme of great importance. The process of information held in the genes of DNA being transcribed into RNA and then translated into protein constitutes the central dogma of genetics. You will want to remember later, that it is the signal of Gibberellic acid that initiates this whole chain of events.

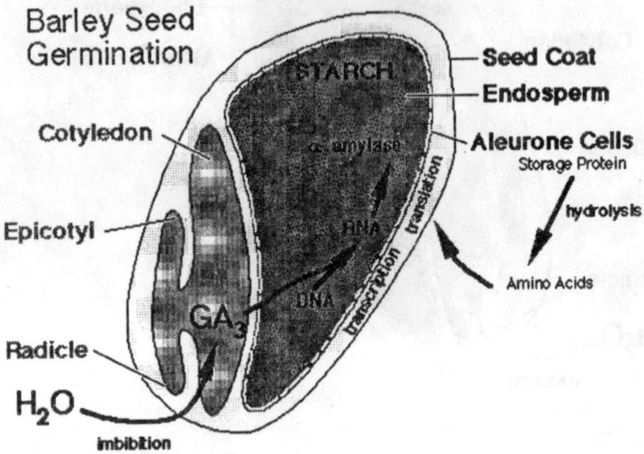

A question one might ask right away is: from where do the amino acid building blocks for the amylase come?

The answer is: from some other biochemistry in the aleurone cells. This biochemistry causes the storage proteins in the aleurone cells to be digested by hydrolytic enzymes. The hydrolysis is accelerated by enzymes known as proteases. These enzymes increase the rate at which the storage protein is cut into individual amino acids.

The amino acids released by the hydrolysis are then free to be reassembled by the ribosomes into the structure of amylase.

The same thing happens in people. You are not what you eat! You do not slowly become a steer by eating beef. The protein in your hamburger is digested into amino acids. Those amino acids are then reassembled into human proteins. Since the instructions for reassembling the amino acids come from your human DNA, the proteins produced are human, not bovine!

In a similar way, barley aleurone storage proteins are digested and amylase is made from the released amino acids.

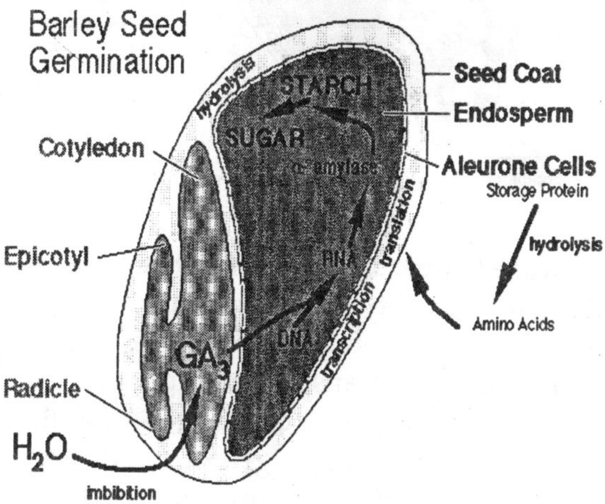

The amylase is secreted (transported out) from the aleurone cells into the endosperm.

Amylase is not just any old protein. It happens to be an enzyme. It catalyzes (accelerates) a particular chemical reaction. Specifically, the amylase speeds the hydrolysis of starch into its component sugar units.

The physiologists now know how the sugar is made in a seed. They can also tell the brewers how to increase the sugar yields for beer making. Note that Gibberellic Acid was the first plant hormone purified in large quantities to be successfully used to increase the yield of any economically-important biochemical in plants!

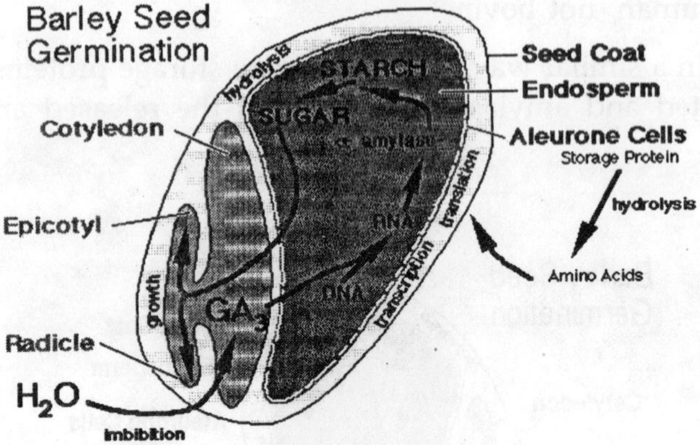

After the released sugar is transported from the endosperm to the embryo. The cotyledon is a major sugar transfer area in grains. Active sugar uptake is its specialized function. The sugar is transported into the embryo and used as a fuel and building block for growth of the embryo itself. The resulting embryo growth will include the emergence of the radicle from the seed/fruit coat. Germination is thus complete.

The next great finding in the study of seed germination is what happens in lettuce seeds (*Lactuca sativa*). Lettuce seeds look somewhat different from barley seeds, because they are dicots. They also have separate seed coat and fruit walls. But the things you buy as lettuce "seeds" are really lettuce fruits (called cypselas). As different as they are, the seed germination process in lettuce needs just one step more than barley to get its seeds germinating.

Lettuce cypselas have very thin fruit walls and seed coats, so light can penetrate these coverings. As is typical of small seeds, light is needed to stimulate seed germination. One might wonder how this would fit into what we just found out with barley.

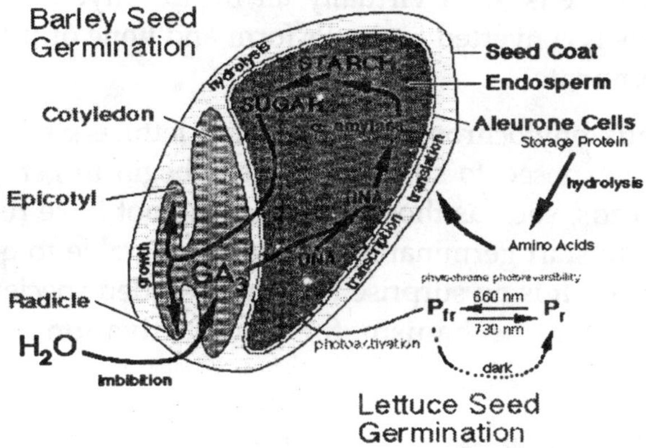

Notice the new step in the diagram. The DNA must first be photoactivated before GA can have its effect. This is accomplished by the formation of a chemical signal known as Pfr.

Pfr is one of two possible chemical forms of a molecule known as phytochrome. Phytochrome is a pigment that absorbs light energy. When it does absorb light energy its internal chemical structure is altered instantly. Thus Pfr can be converted to the alternate Pr form of phytochrome by simply hitting the seeds with a flash of far-red light (730 nm). Similarly, the Pr form of phytochrome can be converted back into Pfr with a flash of red light (660 nm).

Dry lettuce seeds have slightly more than half of their phytochrome in the Pfr form. So if they are moistened and kept in the dark, some of the seeds will actually germinate. If we shine red (or it turns out even white) light on imbibing seeds, then virtually all of the phytochrome is converted to the Pfr form and this photoactivates the genes in the DNA. Almost all of the seeds will germinate in this light!

On the other hand, if we shine far-red light on imbibing seeds, then virtually all of the phytochrome in the seeds is converted to the Pr form and none of the seeds will germinate.

Thus phytochrome and light tell lettuce seeds when they are exposed to sunlight and can begin to germinate. Small seeds, such as those in lettuce, do not have reserves enough to start germinating without being able to quickly get to light. It is no surprise that small-seeded species have evolved this mechanism for photoactivation of seed germination.

Plant Tissue Culture

Most methods of plant transformation applied to GM crops require that a whole plant is regenerated from isolated plant cells or tissue which have been genetically transformed. This regeneration is conducted in vitro so that the environment and growth medium can be manipulated to ensure a high frequency of regeneration. In addition to a high frequency of regeneration, the regenerable cells must be accessible to gene transfer by whatever technique is chosen.

The primary aim is therefore to produce, as easily and as quickly as possible, a large number of regenerable cells that are accessible to gene transfer. The subsequent regeneration step is often the most difficult step in plant transformation studies. However, it is important to remember that a high frequency of regeneration does not necessarily correlate with high transformation efficiency.

Practically any plant transformation experiment relies at some point on tissue culture. There are some exceptions to this generalisation, but the ability to regenerate plants from isolated cells or tissues in vitro underpins most plant transformation systems.

Plasticity and Totipotency

Two concepts, plasticity and totipotency, are central to understanding plant cell culture and regeneration. Plants, due to their sessile nature and long life span, have developed a greater ability to endure extreme conditions and predation than have animals.

Many of the processes involved in plant growth and development adapt to environmental conditions. This plasticity allows plants to alter their metabolism, growth and development to best suit their environment. Particularly important aspects of this adaptation, as far as plant tissue culture and regeneration are concerned, are the abilities to initiate cell division from almost any tissue of the plant and to regenerate lost organs or undergo different developmental pathways in response to particular stimuli.

When plant cells and tissues are cultured in vitro they generally exhibit a very high degree of plasticity, which allows one type of tissue or organ to be initiated from another type. In this way, whole plants can be subsequently regenerated. This regeneration of whole organisms depends upon the concept that all plant cells can, given the correct stimuli, express the total genetic potential of the parent plant. This maintenance of genetic potential is called 'totipotency'. Plant cell culture and regeneration do, in fact, provide the most compelling evidence for totipotency.

In practical terms though, identifying the culture conditions and stimuli required to manifest this totipotency can be extremely difficult and it is still a largely empirical process.

When cultured in vitro, all the needs, both chemical (Table 1) and physical, of the plant cells have to met by the culture vessel, the growth medium and the external environment (light, temperature, etc.). The growth medium has to supply all the essential mineral ions required for growth and development. In many cases (as the biosynthetic capability of cells cultured in vitro may not replicate that of the parent plant), it must also supply additional organic supplements such as amino acids and vitamins.

Table 1. Some of the elements important for plant nutrition and their physiological function.

Element	Function
Nitrogen	Component of proteins, nucleic acids and some coenzymes Element required in greatest amount
Potassium	Regulates osmotic potential, principal inorganic cation
Calcium	Cell wall synthesis, membrane function, cell signalling
Magnesium	Enzyme cofactor, component of chlorophyll
Phosphorus	Component of nucleic acids, energy transfer, component of intermediates in respiration and photosynthesis
Sulphur	Component of some amino acids (methionine, cysteine) and some cofactors
Chlorine	Required for photosynthesis
Iron	Electron transfer as a component of cytochromes
Manganese	Enzyme cofactor
Cobalt	Component of some vitamins
Copper	Enzyme cofactor, electron-transfer reactions
Zinc	Enzyme cofactor, chlorophyll biosynthesis
Molybdenum	Enzyme cofactor, component of nitrate reductase

Many plant cell cultures, as they are not photosynthetic, also require the addition of a fixed carbon source in the form of a sugar (most often sucrose). One other vital component that must also be supplied is water, the principal biological solvent. Physical factors, such as temperature, pH, the gaseous environment, light (quality and duration) and osmotic pressure, also have to be maintained within acceptable limits.

Plant Cell Culture Media

Culture media used for the in vitro cultivation of plant cells are composed of three basic components:

(1) essential elements, or mineral ions, supplied as a complex mixture of salts;

(2) an organic supplement supplying vitamins and/or amino acids; and

(3) a source of fixed carbon; usually supplied as the sugar sucrose.

For practical purposes, the essential elements are further divided into the following categories:

(1) macroelements (or macronutrients);

(2) microelements (or micronutrients); and

(3) an iron source.

Complete, plant cell culture medium is usually made by combining several different components, as outlined in Table 2.

Table 2. Composition of a typical plant culture medium. The medium described here is that of Murashige and Skoog (MS)[a]

Essential element	Concentration in stock solution (mg/l)	Concentration in medium (mg/l)
Macroelements[b]		
NH_4NO_3	33000	1,650
KNO_3	38000	1,900
$CaCl_2.2H_2O$	8800	440
$MgSO_4.7H_2O$	7400	370
KH_2PO_4	3400	170

Microelements[c]		
KI	166	0.83
H_3BO_3	1240	6.2
$MnSO_4.4H_2O$	4460	22.3
$ZnSO_4.7H_2O$	1720	8.6
$Na_2MoO_4.2H_2O$	50	0.25
$CuSO_4.5H_2O$	5	0.025
$CoCl_2.6H_2O$	5	0.025
Iron source[c]		
$FeSO_4.7H_2O$	5560	27.8
$Na_2EDTA._2H_2O$	7460	37.3
Organic supplement[c]		
Myoinositol	20000	100
Nicotinic acid	100	0.5
Pyridoxine-HCl	100	0.5
Thiamine-HCl	100	0.5
Glycine	400	2
Carbon source[d]		
Sucrose	Added as solid	30,000

[a]Many other commonly used plant culture media (such as amborg's B5 and Schenk and Hildebrandt (SH) medium) are similar in composition to MS medium and can be thought of as 'high-salt' media. MS is an extremely widely used medium and forms the basis for many other media formulations.
[b] 50 ml of stock solution used per litre of medium.
[c]5 ml of stock solution used per litre of medium.
[d] Added as solid.

Media Components

It is useful to briefly consider some of the individual components of the stock solutions.

Macroelements

As is implied by the name, the stock solution supplies

those elements required in large amounts for plant growth and development. Nitrogen, phosphorus, potassium, magnesium, calcium and sulphur (and carbon, which is added separately) are usually regarded as macroelements. These elements usually comprise at least 0.1% of the dry weight of plants.

Nitrogen is most commonly supplied as a mixture of nitrate ions (from the KNO_3) and ammonium ions (from the NH_4NO_3). Theoretically, there is an advantage in supplying nitrogen in the form of ammonium ions, as nitrogen must be in the reduced form to be incorporated into macromolecules. Nitrate ions therefore need to be reduced before incorporation. However, at high concentrations, ammonium ions can be toxic to plant cell cultures and uptake of ammonium ions from the medium causes acidification of the medium. In order to use ammonium ions as the sole nitrogen source, the medium needs to be buffered.

High concentrations of ammonium ions can also cause culture problems by increasing the frequency of vitrification (the culture appears pale and 'glassy' and is usually unsuitable for further culture). Using a mixture of nitrate and ammonium ions has the advantage of weakly buffering the medium as the uptake of nitrate ions causes OH- ions to be excreted. Phosphorus is usually supplied as the phosphate ion of ammonium, sodium or potassium salts. High concentrations of phosphate can lead to the precipitation of medium elements as insoluble phosphates.

Microelements

These elements are required in trace amounts for plant growth and development, and have many and diverse roles. Manganese, iodine, copper, cobalt, boron, molybdenum, iron and zinc usually comprise the microelements, although other elements such as nickel and aluminium are frequently found in some formulations.

Iron is usually added as iron sulphate, although iron citrate can also be used. Ethylenediaminetetraacetic acid (EDTA) is usually used in conjunction with the iron sulphate. The EDTA complexes with the iron so as to allow the slow and continuous release of iron into the medium. Uncomplexed iron can precipitate out of the medium as ferric oxide.

Organic Supplements

Only two vitamins, thiamine (vitamin B1) and myoinositol (considered a B vitamin) are considered essential for the culture of plant cells in vitro. However, other vitamins are often added to plant cell culture media for historical reasons.

Amino acids are also commonly included in the organic supplement. The most frequently used is glycine (arginine, asparagine, aspartic acid, alanine, glutamic acid, glutamine and proline are also used), but in many cases its inclusion is not essential. Amino acids provide a source of reduced nitrogen and, like ammonium ions, uptake causes acidification of the medium. Casein hydrolysate can be used as a relatively cheap source of a mix of amino acids.

Carbon Source

Sucrose is cheap, easily available, readily assimilated and relatively stable and is therefore the most commonly used carbon source. Other carbohydrates (such as glucose, maltose, galactose and sorbitol) can also be used, and in specialised circumstances may prove superior to sucrose.

Gelling Agents

Media for plant cell culture in vitro can be used in either liquid or 'solid' forms, depending on the type of culture being grown. For any culture types that require the plant cells or tissues to be grown on the surface of the medium, it must be solidified (more correctly termed 'gelled'). Agar, produced from seaweed, is the most common type of gelling agent, and is ideal for routine applications. However, because it is a natural product, the agar quality can vary from supplier to supplier and from batch to batch. For more demanding applications, a range of purer (and in some cases, considerably more expensive) gelling agents are available. Purified agar or agarose can be used, as can a variety of gellan gums.

Plant Growth Regulators

The essential point as far as plant cell culture is concerned is that, due to this plasticity and totipotency, specific media manipulations can be used to direct the development of plant cells in culture. Plant growth regulators are the critical media components in determining the developmental pathway of the plant cells. The plant growth regulators used most commonly are plant hormones or their synthetic analogues.

Classes of Plant Growth Regulators

There are five main classes of plant growth regulator used in plant cell culture, namely:

(1) auxins;

(2) cytokinins;

(3) gibberellins;

(4) abscisic acid;

(5) ethylene.

Each class of plant growth regulator will be briefly looked at.

Auxins

Auxins promote both cell division and cell growth The most important naturally occurring auxin is IAA (indole-3-acetic acid), but its use in plant cell culture media is limited because it is unstable to both heat and light. Occasionally, amino acid conjugates of IAA (such as indole-acetyl-L-alanine and indole-acetyl-L-glycine), which are more stable, are used to partially alleviate the problems associated with the use of IAA. It is more common, though, to use stable chemical analogues of IAA as a source of auxin in plant cell culture media. 2,4-Dichlorophenoxyacetic acid (2,4-D) is the most commonly used auxin and is extremely effective in most circumstances. Other auxins are available (Table 3), and some may be more effective or 'potent' than 2,4-D in some instances.

Table 3. Commonly used auxins, their abbreviation and chemical name

Abbreviation/name	Chemical name
2,4-D	2,4-dichlorophenoxyacetic acid
2,4,5-T	2,4,5-trichlorophenoxyacetic acid
Dicamba	2-methoxy-3,6-dichlorobenzoic acid
IAA	Indole-3-acetic acid
IBA	Indole-3-butyric acid
MCPA	2-methyl-4-chlorophenoxyacetic acid
NAA	1-naphthylacetic acid
NOA	2-naphthyloxyacetic acid
Picloram	4-amino-2,5,6-trichloropicolinic acid

Cytokinins

Cytokinins promote cell division. Naturally occurring cytokinins are a large group of structurally related (they are purine derivatives) compounds. Of the naturally occurring cytokinins, two have some use in plant tissue culture media (Table 4). These are zeatin and 2iP (2-

isopentyl adenine). Their use is not widespread as they are expensive (particularly zeatin) and relatively unstable. The synthetic analogues, kinetin and BAP (benzylaminopurine), are therefore used more frequently. Non-purine-based chemicals, such as substituted phenylureas, are also used as cytokinins in plant cell culture media. These substituted phenylureas can also substitute for auxin in some culture systems.

Table 4. Commonly used cytokinins, their abbreviation and chemical name

Abbreviation/name	Chemical name
BAP[a]	6-benzylaminopurine
2iP (IPA)[b]	[N6-(2-isopentyl)adenine]
Kinetin[a]	6-furfurylaminopurine
Thidiazuronc	1-phenyl-3-(1,2,3-thiadiazol-5-yl)urea
Zeatin[b]	4-hydroxy-3-methyl-trans-2-butenylaminopurine

[a] Synthetic analogues.
[b] Naturally occurring cytokinins.
[c] A substituted phenylurea-type cytokinin.

Gibberellins

There are numerous, naturally occurring, structurally related compounds termed 'gibberellins'. They are involved in regulating cell elongation, and are agronomically important in determining plant height and fruit-set. Only a few of the gibberellins are used in plant tissue culture media, GA3 being the most common.

Abscisic acid

Abscisic acid (ABA) inhibits cell division. It is most commonly used in plant tissue culture to promote distinct developmental pathways such as somatic embryogenesis.

Ethylene

Ethylene is a gaseous, naturally occurring, plant growth regulator most commonly associated with controlling fruit ripening in climacteric fruits, and its use in plant tissue culture is not widespread. It does, though, present a particular problem for plant tissue culture. Some plant cell cultures produce ethylene, which, if it builds up sufficiently, can inhibit the growth and development of the culture. The type of culture vessel used and its means of closure affect the gaseous exchange between the culture vessel and the outside atmosphere and thus the levels of ethylene present in the culture.

Generalisations about plant growth regulators and their use in plant cell culture media have been developed from initial observations made in the 1950s. There is, however, some considerable difficulty in predicting the effects of plant growth regulators: this is because of the great differences in culture response between species, cultivars and even plants of the same cultivar grown under different conditions.

Figure 1. The effect of different ratios of auxin to cytokinin on the growth and morphogenesis of callus. High auxin to cytokinin ratios promote root development, low ratios promote shoot development. Intermediate ratios promote continued growth of the callus without differentiation.

However, some principles do hold true and have become the paradigm on which most plant tissue culture regimes are based. Auxins and cytokinins are the most widely used plant growth regulators in plant tissue culture and are usually used together, the ratio of the auxin to the cytokinin determining the type of culture established or regenerated (Figure 1). A high auxin to cytokinin ratio generally favours root formation, whereas a high cytokinin to auxin ratio favours shoot formation. An intermediate ratio favours callus production.

Types of Cultures

Cultures are generally initiated from sterile pieces of a whole plant. These pieces are termed 'explants', and may consist of pieces of organs, such as leaves or roots, or may be specific cell types, such as pollen or endosperm. Many features of the explant are known to affect the efficiency of culture initiation. Generally, younger, more rapidly growing tissue (or tissue at an early stage of development) is most effective.

Several different culture types most commonly used in plant transformation studies will now be examined in more detail.

Callus

Explants, when cultured on the appropriate medium, usually with both an auxin and a cytokinin, can give rise to an unorganised, growing and dividing mass of cells. It is thought that any plant tissue can be used as an explant, if the correct conditions are found. In culture, this proliferation can be maintained more or less indefinitely, provided that the callus is subcultured on to fresh medium periodically. During callus formation there is some degree of dedifferentiation (i.e. the changes that occur during development and specialisation are, to some extent, reversed), both in morphology (callus is usually composed of unspecialised parenchyma cells) and metabolism.

One major consequence of this dedifferentiation is that most plant cultures lose the ability to photosynthesise. This has important consequences for the culture of callus tissue, as the metabolic profile will probably not match that of the donor plant. This necessitates the addition of other components-such as vitamins and, most importantly, a carbon source-to the culture medium, in addition to the usual mineral nutrients.

Callus culture is often performed in the dark (the lack of photosynthetic capability being no drawback) as light can encourage differentiation of the callus.

During long-term culture, the culture may lose the requirement for auxin and/or cytokinin. This process, known as 'habituation', is common in callus cultures from some plant species (such as sugar beet). Callus cultures are extremely important in plant biotechnology. Manipulation of the auxin to cytokinin ratio in the medium can lead to the development of shoots, roots or somatic embryos from which whole plants can subsequently be produced. Callus cultures can also be used to initiate cell suspensions, which are used in a variety of ways in plant transformation studies.

Cell-suspension cultures

Callus cultures, broadly speaking, fall into one of two categories: compact or friable. In compact callus the cells are densely aggregated, whereas in friable callus the cells are only loosely associated with each other and the callus becomes soft and breaks apart easily. Friable callus provides the inoculum to form cell-suspension cultures. Explants from some plant species or particular cell types tend not to form friable callus, making cell-suspension initiation a difficult task. The friability of callus can sometimes be improved by manipulating the medium components or by repeated subculturing.

The friability of the callus can also sometimes be improved by culturing it on 'semi-solid' medium (medium with a low concentration of gelling agent). When friable callus is placed into a liquid medium (usually the same composition as the solid medium used for the callus culture) and then agitated, single cells and/or small clumps of cells are released into the medium. Under the correct conditions, these released cells continue to grow and divide, eventually producing a cell-suspension culture.

A relatively large inoculum should be used when initiating cell suspensions so that the released cell numbers build up quickly. The inoculum should not be too large though, as toxic products released from damaged or stressed cells can build up to lethal levels. Large cell clumps can be removed during subculture of the cell suspension. Cell suspensions can be maintained relatively simply as batch cultures in conical flasks. They are continually cultured by repeated subculturing into fresh medium. This results in dilution of the suspension and the initiation of another batch growth cycle. The degree of dilution during subculture should be determined empirically for each culture. Too great a degree of dilution will result in a greatly extended lag period or, in extreme cases, death of the transferred cells.

After subculture, the cells divide and the biomass of the culture increases in a characteristic fashion, until nutrients in the medium are exhausted and/or toxic by-products build up to inhibitory levels-this is called the 'stationary phase'. If cells are left in the stationary phase for too long, they will die and the culture will be lost. Therefore, cells should be transferred as they enter the stationary phase. It is therefore important that the batch growth-cycle parameters are determined for each cell-suspension culture.

Protoplasts

Protoplasts are plant cells with the cell wall removed. Protoplasts are most commonly isolated from either leaf mesophyll cells or cell suspensions, although other sources can be used to advantage. Two general approaches to removing the cell wall (a difficult task without damaging the protoplast) can be taken-mechanical or enzymatic isolation.

Mechanical isolation, although possible, often results in low yields, poor quality and poor performance in culture due to substances released from damaged cells. Enzymatic isolation is usually carried out in a simple salt solution with a high osmoticum, plus the cell wall degrading enzymes. It is usual to use a mix of both cellulase and pectinase enzymes, which must be of high quality and purity.

Protoplasts are fragile and easily damaged, and therefore must be cultured carefully. Liquid medium is not agitated and a high osmotic potential is maintained, at least in the initial stages. The liquid medium must be shallow enough to allow aeration in the absence of agitation. Protoplasts can be plated out on to solid medium and callus produced. Whole plants can be regenerated by organogenesis or somatic embryogenesis from this callus. Protoplasts are ideal targets for transformation by a variety of means.

Root cultures

Root cultures can be established in vitro from explants of the root tip of either primary or lateral roots and can be cultured on fairly simple media. The growth of roots in vitro is potentially unlimited, as roots are indeterminate organs. Although the establishment of root cultures was one of the first achievements of modern plant tissue culture, they are not widely used in plant transformation studies.

Shoot tip and meristem culture

The tips of shoots (which contain the shoot apical meristem) can be cultured in vitro, producing clumps of shoots from either axillary or adventitious buds. This method can be used for clonal propagation. Shoot meristem cultures are potential alternatives to the more commonly used methods for cereal regeneration as they are less genotype-dependent and more efficient (seedlings can be used as donor material).

Embryo culture

Embryos can be used as explants to generate callus cultures or somatic embryos. Both immature and mature embryos can be used as explants. Immature, embryo-derived embryogenic callus is the most popular method of monocot plant regeneration.

Microspore culture

Haploid tissue can be cultured in vitro by using pollen or anthers as an explant. Pollen contains the male gametophyte, which is termed the 'microspore'. Both callus and embryos can be produced from pollen. Two main approaches can be taken to produce in vitro cultures from haploid tissue. The first method depends on using the anther as the explant. Anthers (somatic tissue that surrounds and contains the pollen) can be cultured on solid medium (agar should not be used to solidify the medium as it contains inhibitory substances).

Pollen-derived embryos are subsequently produced via dehiscence of the mature anthers. The dehiscence of the anther depends both on its isolation at the correct stage and on the correct culture conditions. In some species, the reliance on natural dehiscence can be circumvented by cutting the wall of the anther, although this does, of course, take a considerable amount of time. Anthers can also be cultured in liquid medium, and pollen released from the anthers can be induced to form embryos, although the effi-ciency of plant regeneration is often very low. Immature pollen can also be extracted from developing anthers and cultured directly, although this is a very time-consuming process.

Both methods have advantages and disadvantages. Some beneficial effects to the culture are observed when anthers are used as the explant material. There is, however, the danger that some of the embryos produced from anther culture will originate from the somatic anther tissue rather than the haploid microspore cells. If isolated pollen is used there is no danger of mixed embryo formation, but the efficiency is low and the process is time-consuming. In microspore culture, the condition of the donor plant is of critical importance, as is the timing of isolation.

Pretreatments, such as a cold treatment, are often found to increase the efficiency. These pretreatments can be applied before culture, or, in some species, after placing the anthers in culture. Plant species can be divided into two groups, depending on whether they require the addition of plant growth regulators to the medium for pollen/anther culture; those that do also often require organic supplements, e.g. amino acids.

Many of the cereals (rice, wheat, barley and maize) require medium supplemented with plant growth regulators for pollen/anther culture. Regeneration from microspore explants can be obtained by direct embryogenesis, or via a callus stage and subsequent embryogenesis. Haploid tissue cultures can also be initiated from the female gametophyte (the ovule). In some cases, this is a more efficient method than using pollen or anthers.

The ploidy of the plants obtained from haploid cultures may not be haploid. This can be a consequence of chromosome doubling during the culture period. Chromosome doubling (which often has to be induced by treatment with chemicals such as colchicine) may be an advantage, as in many cases haploid plants are not the desired outcome of regeneration from haploid tissues. Such plants are often referred to as 'di-haploids', because they contain two copies of the same haploid genome.

Plant Regeneration

Having looked at the main types of plant culture that can be established in vitro, we can now look at how whole plants can be regenerated from these cultures. In broad terms, two methods of plant regeneration are widely used in plant transformation studies, i.e. somatic embryogenesis and organogenesis.

Somatic Embryogenesis

In somatic (asexual) embryogenesis, embryo-like structures, which can develop into whole plants in a way analogous to zygotic embryos, are formed from somatic tissues (Figure 2). These somatic embryos can be produced either directly or indirectly. In direct somatic embryogenesis, the embryo is formed directly from a cell or small group of cells without the production of an intervening callus. Though common from some tissues (usually reproductive tissues such as the nucellus, styles or pollen), direct somatic embryogenesis is generally rare in comparison with indirect somatic embryogenesis.

In indirect somatic embryogenesis, callus is first produced from the explant. Embryos can then be produced from the callus tissue or from a cell suspension produced from that callus. Somatic embryogenesis usually proceeds in two distinct stages. In the initial stage (embryo initiation), a high concentration of 2,4-D is used. In the second stage (embryo production) embryos are produced in a medium with no or very low levels of 2,4-D.

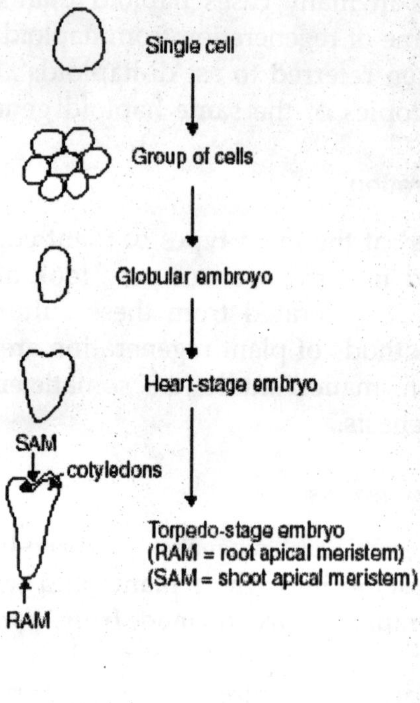

Figure 2. A schematic representation of the sequential stages of somatic embryo development

In many systems it has been found that somatic embryogenesis is improved by supplying a source of reduced nitrogen, such as specific amino acids or casein hydrolysate.

The principal method adopted for the tissue culture and regeneration of a wide range of cereal species is somatic embryogenesis, using cultures initiated from immature zygotic embryos. Embryogenic callus is normally initiated by placing the immature embryo on to a medium containing 2,4-D. Shoot regeneration is initiated by placing the embryogenic callus on a medium with BAP (with or without 2,4-D). These shoots can be subsequently rooted. The medium used for induction of embryogenic callus is usually a modi.ed MS (for Triticeae) or N6 (for rice and maize). Maltose is often used as the carbon source in preference to sucrose, and additional organic supplements (such as speci.c amino acids, yeast extract and/or casein hydrolysate) are common.

The isolation and culture of immature embryos is, however, a labour-intensive and relatively expensive procedure. An additional problem is the small target size if the immature embryos are to used for biolistic transformation. Alternatives are therefore being sought. One alternative is to use mature embryos (or seeds) as the explant to initiate embryogenic callus. This approach has been successfully applied to several cereal species such as rice and oats. The culture techniques and media used for culture establishment and regeneration from mature embryo/seed-derived cultures are fundamentally the same as those used for cultures initiated from immature embryos.

Importance of genotype

The major influence on tissue-culture response appears to be genetic, with culture requirements varying between species and cultivars. Model genotypes that responded well to culture in vitro were initially used in plant transformation studies. However, most of the model genotypes used were not elite, commercial cultivars. The commercial cultivars tended to respond poorly to culture in vitro. One of the main aims is therefore to identify the components that make up a widely applicable, optimal culture regime.

Many factors have been investigated for their ability to improve the culture response from elite cultivars, including media components (such as alternative carbon sources, macro- and microelement concentrations and composition), media preparation method and donor plant condition and growth conditions.

Organogenesis

Somatic embryogenesis relies on plant regeneration through a process analogous to zygotic embryo germination. Organogenesis relies on the production of organs, either directly from an explant or from a callus culture. There are three methods of plant regeneration via organogenesis.

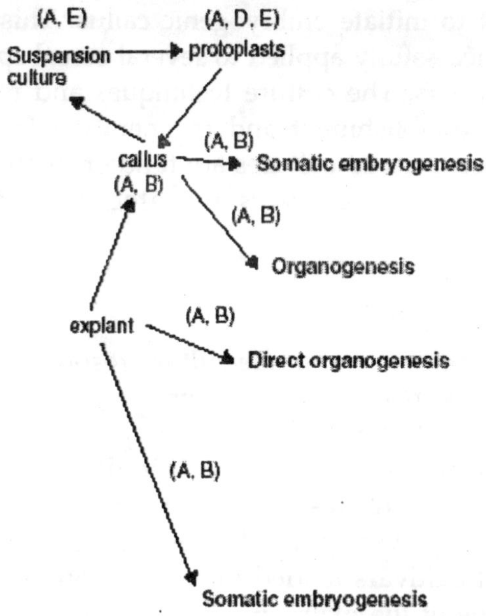

Figure 3. A simplified scheme for the integration of plant tissue culture into plant transformation protocols

The first two methods depend on adventitious organs arising either from a callus culture or directly from an explant (Figure 3). Alternatively, axillary bud formation and growth can also be used to regenerate whole plants from some types of tissue culture.

Organogenesis relies on the inherent plasticity of plant tissues, and is regulated by altering the components of the medium. In particular, it is the auxin to cytokinin ratio of the medium that determines which developmental pathway the regenerating tissue will take. It is usual to induce shoot formation by increasing the cytokinin to auxin ratio of the culture medium. These shoots can then be rooted relatively simply.

Various methods of plant regeneration are available to the plant biotechnologist. Some plant species may be amenable to regeneration by a variety of methods, but some may only be regenerated by one method. Not all plant tissue is suited to every plant transformation method, and not all plant species can be regenerated by every method. There is therefore a need to find both a suitable plant tissue culture/regeneration regime and a compatible plant transformation methodology.

9

Applicaiton of IPR in Biotechnology

One of the most significant developments associated with the advent of biotechnology has been the strengthening of intellectual property protection for biological inventions. Developments in biotechnology and the recent commercial valuation of genetic resources are putting pressure on governments around the world to ensure the protection of intellectual property rights.

Tangible property rights derive from the land and anything found on it including plants, animals, microorganisms and their genes. Intellectual property rights, on the other hand, arise from the intangible information generated from genetic resources. They are the legal instruments which confer protection on processes or products of research and development efforts and formally assure the allocation of benefits to the innovator in return for full disclosure to society.

The rationale for their existence is that they protect useful scientific advances and thereby create incentives for investment in future developments. Intellectual property rights comprise a variety of legal devices such as patents, plant breederís rights, trade secret laws, trademarks and copyrights which afford varying degrees of protection and benefits to the inventor. Increasingly, these rights are being extended to cover biological inventions.

Protection of Biological Inventions

Patents confer the highest level of protection and ensure the greatest profits for the inventor. For an invention to be patented in the United States, it must meet the criteria of novelty, utility, and non-obviousness. Thus, patents may not be granted for naturally occurring species (according to the 'product of nature' doctrine). In this regard, the US Plant Patent Act of 1930 was significant in that it allowed, for the first time, patent protection for certain plants varieties.

The 1980 *Diamond* vs. *Chakrabarty* decision marked the first instance where a patent was granted for the development of a novel genetically engineered microorganism. It was ruled that anything under the sun that was created by man was patentable, thereby ending the general hesitation to extend patent protection to biological inventions.

Subsequently, there has been a boom in patent claims filed with the US Patent and Trademark Office. Microorganism patents are now obtainable in most industrialised countries. Court rulings have incrementally brought plant varieties, microorganism and animals under the patent system. The first transgenic animal patent was granted to Harvard University for the onco-mouse in 1988. Such patents are under criticism in Europe because they are deemed by human rights organisations to be antithetical to morality.

The ethics of the ownership of living entities has most recently been brought to the forefront of the intellectual property right debate with the submission of patent applications for animal-human hybrids. At issue is the degree to which a creature is human enough to preclude patentability. A recent decision has rejected the patent application on the grounds that there is precedence for the scientific methods cited and that the Congress of the United States did not intend humans to be patentable. The case illustrates how patent offices may serve as sites where social and ethical issues surrounding the protection of living entities by property rights are debated.

The advent of plant genome sequencing efforts has resulted in the patenting of plant DNA sequences by the plant biotechnology industry and public research institutions. DNA sequences are patentable only when the gene has been isolated and a utility for it demonstrated. In this case, the current intellectual property rights regime has largely adapted the existing case law for chemical compounds to the patenting of DNA sequences. Such patents cover the process of isolating and discovering a particular DNA segment rather than the unmodified sequence itself. The issue of patenting fragments of gene sequences or expressed sequence tags (ESTs) is a particularly contentious one as they are often not associated with any function.

However, the US PTO has indicated that they are patentable if they could be used as molecular markers or probes to identify specific genetic sequences. The patenting of single nucleotide polymorphisms (SNPs) is also currently being pursued by the biotechnology industry. Critics argue, however, that while SNPs may be useful tools in the discovery of genes, they fail the test of novelty and, therefore, should not be patentable.

Harmonisation of IPR regimes

While the US pursued patent protection of certain plant varieties, other countries worked through the International Union for the Protection of New Varieties of Plants (UPOV) created in 1961 to afford protection by granting plant breedersí rights to certain plant varieties. An exemption is the 'farmers' privilege', which allows member states the option to allow farmers to save seed for their own use.

The 1991 amendment extends plant breedersí rights to all plant genera and species that satisfy the criteria of distinctiveness, utility and stability. Additionally, exclusive rights are provided to the breeder which cover all harvested material and which allow the collection of royalties from the sale of seeds.

At its inception, UPOV did not anticipate the inclusion of plant varieties within the patent system. With the 1991 amendment, for the first time, a plant may now be protected by a patent as well as by UPOV-style plant breedersí rights. The situation highlights the process whereby developments in biotechnology promote the harmonisation of intellectual property regimes.

Currently, the European Directive on the Protection of Biological Inventions allows patents on those plants and animals where the invention may be applied to more than one variety. For instance, wheat modified by the B.t. gene is not considered a plant variety under UPOV but may be patentable.

Efforts to harmonise intellectual property rights regimes in the international arena should promote cooperative efforts between nations. However, calls for the patent-like protection for plants on an international scale by WTO's Trade Related Aspects of Intellectual Property Rights (TRIPs) agreement are deemed to be at odds with the provisions of the Convention on Biological Diversity (CBD).

The TRIPs agreement stipulates that signatories recognise patents on wide variety of processes and products including microorganisms or risk sanctions by the World Trade Organisation (WTO). Although animals are currently exempt, the agreement requires member states to institute protection for plant varieties either via a traditional patent regime or the creation of a sui generis ('of its own kind') property rights framework.

While the revised UPOV model advocates patent-like protection for plant varieties, a sui generis system could presumably allow for other systems for plant variety protection including provisions for 'farmers' rights' and community property rights in keeping with the diverse needs of different countries.

Such an interpretation has been adopted by some countries such as India who would like to create a sui generis system that would formalise the protection of indigenous knowledge, ensure the equitable distribution of benefits, and facilitate the access and transfer of technology. Such a system would be in the spirit of the CBD which recognises the sovereign right of states over their resources and their right to regulate access to such resources.

The case of germination control technology (where seeds become sterile after one planting), for instance, illustrates the conflict between the interests of multinational seed companies which seek to maximise profits and the farming practices of indigenous and local communities. While the biotechnology industry sees intellectual property rights as essential to recoup the costs of research and development, farmers see their traditional practices such as the right to save seed being undermined.

Alternative approaches being developed involve ways of disabling certain traits conferred by genetic modification. Such traits could then be activated by proprietary chemicals which have no influence on the germination process itself.

Public-private Sector Conflicts

The private sector, for instance, holds the dominant position over public institutions in filing patent claims for DNA sequences as well as for patents for basic biotechnology processes ('enabling patents'). Between 1980 and 1996, 48% of the patent claims for DNA sequences derived from 78 plant species of economic or scientific importance were made by 14 multinationals alone, with Monsanto having the largest stake.

This concentration of patents and the trend toward consolidation of the agrochemical and seed industry raises the matter of access to key technologies. Access is likely to be dictated by the degree to which corporations are willing to license key technologies to smaller corporations and developing countries and the cost at which they are willing to do so.

This is of particular concern to developing countries, which may view biotechnology as a potentially fruitful economic sector and yet may not be able to achieve this objective due to high licensing costs. The breadth and nature of certain patents is also seen by some as a matter of concern since it might allow for monopoly suppliers. Regulation of access to genetic resources has been advocated by the CBD and, currently, legislation and policy measures are being discussed in a number of countries.

Classifying Utility Patents into Technology Categories and Subcategories

Agricultural biotechnology is biotechnology applied to agriculture. Most generally, agricultural biotechnology is the use of organisms or parts of an organism to make or improve products or processes in agriculture. The domestication of plant species and selection of desired characteristics within agricultural species would qualify under this definition. More narrowly, what *biotechnology* represents today is new knowledge about the natural processes of DNA replication, breakage, ligation, and repair that has made possible a deeper understanding of the mechanics of cell biology and the hereditary process itself.

Although in agriculture the term "biotechnology" has been most closely associated with genetic manipulation at the DNA level, or *genetic engineering*, it may refer to a variety of techniques and products. These may include, for example, use of molecular markers in genetic improvement or more general use of genomic information. Similarly, the use of enzymes for fermentation in brewing or cheesemaking would be early examples of a broad definition of "food biotechnology." Genetically engineering yeast to modify or improve a baking process would be an example of a narrower, or more recent definition of food biotechnology.

In the complex and constantly changing world of agricultural biotechnology, definition and classification are difficult tasks. The U.S. Patent and Trademark Office (USPTO) has its own classification system, with classes and subclasses, but these are intended to encompass all utility patents. "Agricultural Biotechnology" patents can be found in several major USPTO categories or some of their subcategories, including 800: Multicellular Living Organisms And Unmodified Parts Thereof And Related Processes; 935: Biotechnology (a now-abandoned category); 435: Chemistry: Molecular Biology and Microbiology; 47: Plant Husbandry; 424 and 514: Drug, Bio-Affecting and Body Treating Compositions; and 504: Plant Protecting and Regulating Compositions.

Agricultural applications may cut across more than one of these categories. Furthermore, USPTO categories and classifications change with evolving technology. For example, the 935 category, while useful in the filtering process, no longer exists and cannot serve as a guide to complete classification.

Principles of the Classification Scheme

Several general principles governed the development of the ERS classification scheme. First, because defining biotechnology is difficult, we wanted to create a very broad database. Users would then be able to choose patents from different subsets of the data to conform to their own definitions of "agricultural biotechnology."

The broadest definition would result in selecting all the utility patents in our database, although of course the filtering process as well as conceptual difficulties undoubtedly excluded patents relevant for some users. On the other hand, a focus on genetic engineering in agricultural plants could be served by selecting patents in several categories relating to genetically engineered plants, transformation technologies, and so on.

Second, from the beginning we tried to distinguish biotechnology products (e.g., new crop varieties genetically engineered to resist pests) from research tools or enabling technologies (e.g. genetic transformation through micro-projectile bombardment or selectable marker genes).

Third, as with the USPTO classification scheme or the international classification scheme, categories were expected to overlap. For example, a patent might be classified under "genomics" because it covers identification of a beneficial plant genetic trait with genomic sequencing of its DNA; under "genetic transformation" of a plant to include this trait; and under "protection, nutrition, and biological control of plants and animals" because the transformed plant expresses an improved agronomic property such as higher yield, pest resistance, drought tolerance, etc.

The classification scheme was first developed before we populated the database through the filtering process. ERS researchers began by drafting a classification scheme, based on consultation with several major sources. After this, scientists with experience in agricultural biotechnology reviewed the scheme, in particular one of our IFAFS (Initiative for Future Agriculture and Food Systems) collaborators at Rutgers University, a molecular biochemist, who reviewed the classification system on several occasions. This resulted in additions and modifications to the scheme initially proposed. This process proceeded through several iterations.

In addition to the major input received from our agricultural biotechnology research collaborator, four major sources contributed to the classification scheme:

(1) the USPTO classification scheme, with additional information provided by personal communication with the biotechnology art group within the USPTO;

(2) the biotechnology classes used by the Derwent World Patents Index, a private sector subscription service that provides information on patents granted in more than 40 jurisdictions worldwide;

(3) a classification scheme developed by biotechnology specialists at ISNAR (International Service for National Agricultural Research); and

(4) a classification system developed by University of California-Berkeley researchers, after they had used proprietary software developed by Aurigin, Inc., to group agricultural biotechnology patents based on use of similar terms.

As noted, classifying agricultural biotechnology patents is complex. The process described above at one point resulted in a scheme with 15 main technology groupings and 206 sub-technologies. In an attempt to streamline the scheme, make it more user-friendly for a web-based application, and create a scheme in which most of the subtechnologies would be populated with patents, we simplified the classification system by combining or dropping subtechnologies, and combining main technology groups.

Two basic approaches to determining whether any individual patent might fall into one or more technology categories were tested. The first would be to examine each patent individually and categorise it. This would have the advantage of eventually classifying every patent in the database, as well as providing an indepth review of "problem" patents. The disadvantage of this approach is the inordinate amount of time that would be necessary to classify all patents in the database into multiple categories. This disadvantage was borne out in a trial run in which we manually classified patents into a single simple technology class.

The second approach would be to write a series of logical rules (i.e., concerning specific words in specific parts of the patent, or USPTO headings and subheadings) that would determine whether a patent would fall into a given category or not. The advantages of this approach are closely linked to the disadvantages of the first approach, and its disadvantages closely related to the advantages of the first approach. Classifying patents with logical rules is considerably faster (although the rules first have to be written and tested), but in many if not all cases, classification might be less accurate. Some patents might be placed into categories in which they did not belong; others might not be classified into patents in which they did belong. In the end, we opted for this approach.

We developed a set of logical rules intended to sort patents into at least one of the technology categories; multiple categories for a single patent were allowed. We then applied these rules to the patents that were populating the database at the time. We took samples of the patents sorted into each major category, and checked the patents in the samples individually to make sure they belonged in the category to which they were assigned. We also took samples of patents that were not sorted to any category and checked the patents in these samples to see which category they might belong in. On the basis of these procedures, we then modified the logical rules for each technological category.

The logical rules may be modified further as we see the impacts of applying the rules as they now stand. For example, out of over 11,000 utility patents currently in our database, about 1,600 currently are not sorted into any category. Obvious next steps would be to modify the rules further so that very few patents in the database remain uncategorised; to check the patents in a given category to make sure they belong; and to apply the rules to patents that were not filtered into the database originally to see if the database should be expanded.

Enforcement of the Plant Variety Protection Act (PVPA)

Enforcement of the Plant Variety Protection Act (PVPA) in small grains has gained a lot of attention the last couple of years. Numerous individuals, including sellers, seed conditioners, and buyers, in Texas, Arkansas, Oklahoma, and other states have been prosecuted for not abiding by the PVPA.

The following should shed some light on the issue.

Plant Variety Protection Act (PVPA) - enacted December 1970 provides legal intellectual property rights protection to developers of new varieties of plants that are sexually reproduced (by seed). The 1970 PVPA provides the Plant Breeders with property rights over new variety releases. Farmers may save seed to plant their own holdings (land owned, rented, or leased) or sell that amount of seed to a neighbor, if plans for that seed change. All seed sales must comply with state laws, including Title V.

1994 Amended PVPA - an amendment to the Plant Variety Protection Act of 1970. The 1994 PVPA prohibits the sale of all farmer-saved seed without the permission of the variety owner. The length of protection under the PVPA act increased to 20 years for most crops, including wheat and oats. It applies to all varieties protected after April 4, 1995.

Major amendments to the PVPA in 1994:

— Prohibits the sale of all seed of a protected variety without the permission of the variety owner.

— Requires protected varieties to be sold by variety name.

— Increased the protection length from 18 to 20 years.

— Title V – an amendment to the Federal Seed Act. Title V allows seed to be sold by variety name only and as a class of certified seed. Non-certified sales are prohibited. Seed certification is based on the requirements of the Texas Department of Agriculture. For practical purposes, Title V also protects most varieties protected by the PVPA.

Utility Patents – a means of protection for certain varieties containing specific genes, usually developed through genetic engineering or biotechnology. Farmers may not save, clean/condition, or sell any seed protected under a utility patent. Examples include Clearfield Wheat, Roundup Ready crops, Bollgard, and others.

Access to new varieties with increased yield potential, higher quality grain, better disease and insect resistance, and tolerance to herbicides play a vital economic role in Texas wheat production. Improved genetics are estimated to account for well over 60 percent of the yield increase that has occurred over the past century. The primary purpose of the Plant Variety Protection Act was to encourage further development of new non-hybrid varieties in crops such as wheat, oats, and other self-pollinating crops.

Allowing Plant Breeders to determine who can sell seed of new varieties provides them with the ability to recoup the monies expended in the variety development process and re-invest in future development. Prior to PVPA, little, if any, financial incentive existed for breeders and seed companies to invest in developing non-hybrid crops. Additionally, passage of the 1994 amended PVPA has allowed the United States to participate in the international plant breeder's rights treaty. As a result, the proprietary rights on varieties are now respected in many countries worldwide.

Nearly all varieties released by private companies since 1970 are protected by the initial PVPA or the 1994 amended PVPA. Most public release varieties (varieties released by universities) are also protected by the PVPA. Variety protection under the PVPA expires 20 years after protection has been granted..

From a practical standpoint, most varieties planted today are protected under the 1994 PVPA, meaning the variety owner's permission is required prior to any seed sales. All varieties protected under the PVPA must be clearly marked on the seed tag or bulk label indicating the type of protection. The seller is responsible for to informing the buyer if a variety is protected. A farmer can save seed protected under both the 1970 and 1994 PVPA for planting his or her own holdings (land owned, rented, or leased). Under the 1970 PVPA, a farmer can sell only the amount needed to plant his own holdings. However, under the 1994 PVPA, no seed can be sold, unless the variety owner grants permission.

Infringement of the owner's rights under the 1994 PVPA include:

— Selling, offering, delivering, consigning, exchanging, or advertising for sale a protected variety.

— Dispensing the variety to another person without informing that person the variety is protected.

— Inducing a third party to commit any of the above acts.

— Selling a PVPA variety as VNS

Any actions taken as a step in marketing farmer-saved seed are infringements of the rights of the variety owner. A third party can clean and/or condition seed for any farmer, as long as the volume of seed cleaned does not exceed the planned planting acreage of the farmer. However, conditioning, storing, or delivering an unreasonable amount of farmer-saved seed for sale is in violation of the PVPA.

It is strongly recommended that anyone conditioning farmer-saved seed obtain written documentation from the farmer stating the seed conditioning is not in violation of the PVP laws or patents. A farmer having seed cleaned and conditioned by a third party is not in violation of the PVPA, as long as the seed volume does not exceed a planned planting acreage.

The owner of the variety has the option to choose Title V protection or not. Title V protection means that sales of non-certified seed of Title V protected varieties is illegal. Title V of the Federal Seed Act also protects most small grains varieties that are protected under the PVPA. The PVPA states that a protected variety must be sold by the variety name only. Sale of a protected variety as VNS is an infringement of the PVPA. For seed to be sold as VNS, the wheat seed must be from a variety not protected by the 1994 PVPA. As mentioned previously, it is the seller's responsibility to inform the buyer if a variety is protected.

The owner of the variety may bring civil action against persons infringing on his or her rights as stated in the 1970 and 1994 PVPA. The damages awarded by a court must at least compensate the variety developer for the infringement but can be up to triple damages, when willful infringement is found. Additionally, violation of any provisions of the Federal Seed Act, including Title V, is a misdemeanor and is punishable by a fine not to exceed $2,000.

India's Plant Variety Protection and Farmers' Rights Act, 2001

The General Agreement on Trade and Tariffs (GATT), the predecessor to the World Trade Organisation (WTO), was established to restore world trade after the end of the Second World War in 1945. Several GATT rounds starting from 1948, dealt with the quotas and duties of tradable commodities between nations. The 1986 GATT Round, popularly known as the Uruguay Round, brought in new elements into the trade discussion, especially relating to agriculture.

One of the most controversial agreements of the Uruguay Round is that relating to the granting of Intellectual Property Rights on biological materials embodied in the Trade Related Intellectual Property Rights (TRIPS) chapter. TRIPS specifically requires member nations to grant patents on microorganisms, non/biological and microbiological processes as well as effective IPR protection for plant varieties. TRIPS provides a choice for protecting plant varieties. Members may choose from patents, a sui generis system or a combination of the two.

Most developing countries including India have decided not to have patents for plant varieties and have chosen the sui generis option instead. The sui generis system (translating roughly into selfgenerating) means any system a country decides on, provided it grants effective Plant Breeders' Rights. TRIPS does not specify what kind of breeders' rights is meant and it does not say what else a member state can include in its law, apart from breeders' rights. In short, TRIPS is a flexible system, which leaves a lot to the discretion of members. As a response to the TRIPS agreement, India has started enacting a series of domestic laws to implement the commitments it has made. The Plant Variety Protection and Farmers Rights Act, 2001, is the Indian sui generis legislation.

The Indian law, which has been hailed as a progressive, pro-developing country legislation, has some notable features. Apart from a well-defined breeders' right, it has strong and proactive farmers' rights. In fact the Indian legislation succeeds in balancing the rights of Breeders and Farmers and exploits the flexibility granted in TRIPS, in an intelligent manner. There are clauses to protect the rights of researchers and provisions to protect the public interest.

The Indian legislation is the first in the world to grant formal rights to farmers in a way that their self-reliance is not jeopardised. What is significant and positive about this legislation is that it charts its own course, deviating from the norms set by the Union for the Protection of New Plant Varieties (UPOV). UPOV is at present the only platform for regulating plant breeders' rights. It is a platform for developed countries which is modulated to protect the interests of agriculture in industrial countries. It does not even have the notion of farmers rights.

The innovative Indian legislation has opened up interesting possibilities for developing a developing country platform for regulating breeders' and farmers' rights so that both, not just one, are acknowledged and protected. The salient features of the new law are described in this article.

Breeders' Rights

Breeders' rights over the varieties they have developed are more than adequately protected by the draft legislation. On registration, the breeder has rights of commercialisation for the registered variety either in his/her own person or through anyone he designates. These rights include the right to produce, sell, market, distribute, import or export a variety, in short, full control over formal marketing.

The strong protection granted to a plant breeder over his/her variety is seen in the section dealing with infringement of breeders' rights where punishment in the form of substantial fines and jail terms has been prescribed for those who infringe the rights of the registered breeder.

Penalties for Infringing Breeders' Rights

Violation of breeders' rights can be construed at several levels. It applies to the variety itself as also to its packaging. Infringement will be established if the packaging is the same or even similar, such that the package could appear to be that of the breeder. Legally, a similar-looking package will be considered 'Passing Off' and hence actionable. Anyone other than the breeder naturally cannot use the registered name or denomination. The use of the same or similar name in any way, by action or even suggestion, will constitute a violation and will be punishable. Penalties are prescribed for applying false denomination and for selling varieties to which false denominations are applied.

The breeders' rights have been strengthened to the extent that if there is mere suspicion of violation or infringement, the onus of proving innocence is placed on the alleged violator. In any prosecution for falsely using a denomination, the burden of proof is reversed and it is incumbent on the alleged violator to prove that the consent of the breeder was obtained. This is somewhat excessive and needs to be toned down. The normal course in law is for the accuser to furnish proof for the accusation and so it must remain in this case too.

Penalties can range from Rs 50,000 to Rs 10 lakh as well as a jail term ranging from three months to two years, depending on the severity of the damage caused. If the violator is actually selling, offering for sale or merely in the possession of a registered variety belonging to someone else, the punishment is somewhat worse. The penalty remains the same, between Rs 50,000 and 10 lakh but the jail term applicable will not be less than six months, going up to two years. If the offence is repeated, the minimum jail term prescribed is one year, extending to three years and the fine starting at Rs one lakh, can go up to Rs 20 lakh.

The Indian legislation, in providing a well-defined breeder's right, provides sufficient incentive for the seed industry to invest in this sector. At the same time, it is important to recognise that IPR protection does not necessarily deliver a successful product. If a variety decisively provides an advantage, it will be bought, if it does not, it will fool the farmers for a few seasons and then fail. It is also necessary to keep in mind that all IPR systems must strike a balance between the monopoly granted to the IPR holder, in this case the plant breeder, and the benefits to society, in this case the farmers and consumers.

Since nobody concerned with public interest would want plant breeding to shift into just a few hands, it is important to maintain competition and vitality in the plant-breeding sector. That is why freedom and rights for other researchers to use all genetic material, including IPR protected material, is important. An IPR system in a country should not grant such strong rights to breeders that farmers suffer and their livelihoods are threatened. On the other hand, the breeder's innovation should be rewarded so that they continue to breed useful varieties to benefit agricultural and food security.

Rights of Researchers

The Bill has provisions for researchers' rights, which allows scientists and breeders to have free access to registered varieties for research. The registered variety can also be used for the purpose of creating other, new varieties. The breeder cannot stop other breeders from using his/her variety to breed new crop varieties except when the registered variety needs to be used repeatedly as a parental line. In that case authorisation is required. There are however some views that the Indian law actually grants very restricted rights to researchers because of the acknowledgement of Essentially Derived Varieties (EDV), which is defined in detail in the 1991 UPOV Convention.

According to the expansive definition of EDVs, it is felt that all kinds of research will become subject to the breeders' authorisation if a protected variety is used for research. In the Indian Act, the breeders' authorisation is needed for making EDVs. The processes for making EDV have been made so allencompassing in UPOV (natural selection, mutant selection, somaclonal variants, backcrosses and transformation by genetic engineering), that all known forms of creating new varieties will be covered. This would squeeze the researcher's space to the extent that for practically any kind of research on the protected variety, the authorisation of the breeders will be needed, establishing their control on a lot of germplasm.

The Patent Second Amendment Act, 2002

The Plant Variety Protection Act is supportive of research on transgenics because of the strong rights it grants to plant breeders. However, it is the Patent Second Amendment Act that makes distinct concessions to the biotechnology sector and the breeding of transgenic crops. According to this Act, process patents will be allowed on microbiological, biochemical and biotechnological processes. In this way, methods of genetic engineering, processes in the pharmaceutical industry using microorganisms and related processes will be patentable.

There is a curious detail in this section on what constitutes a patentable process, which reflects the special adjustments made for the biotechnology sector. Processes and methods for making plants resistant to disease and for increasing their value or the value of their products, will be patentable. This appears to be tailor-made for the Bt cotton situation and other Bt and Bt-like approaches to introduce resistance to disease.

Discoveries, however, will not be patentable under the Indian law. To qualify for a patent, invention will have to be demonstrated. Plants and animals and species of plants and animals have been kept out of the purview of patents, so have plant varieties and seeds. New varieties of crops and their seeds are thus outside the patent system. Though the Indian law permits process patents, this will not apply to the crucial sector of food. Methods and processes of agriculture and horticulture cannot be patented nor can any other biological processes. The Patent Amendment Act also does not allow the patenting of cells, cell lines, cell organelles like mitochondria and genes.

Farmers' Rights

The Act recognises the farmer not just as a cultivator but also as a conserver of the agricultural gene pool and a breeder who has bred several successful varieties. The Act makes provisions for such farmer's varieties to be registered, with the help of NGOs so that they are protected against being scavenged by formal sector breeders. The rights of rural communities are acknowledged as well. Farmers' rights are defined in the following way:

> The farmer . . . 'shall be deemed to be entitled to save, use, sow, resow, exchange, share or sell his farm produce including seed of a variety protected under this Act in the same manner as he was entitled before the coming into force of this Act.;
>
> Provided that the farmer shall not be entitled to sell branded seed of a variety protected under this Act. Explanation: For the purpose of clause (iii) branded seed means any seed put in a package or any other container and labelled in a manner indicating that such seed is of a variety protected under this Act.'

This formulation allows the farmer to sell seed in the way he has always done, with the restriction that this seed cannot be branded with the breeder's registered name. In this way, both farmers' and breeders' rights are protected. The breeder is rewarded for his innovation by having control of the commercial market place but without being able to threaten the farmers' ability to independently engage in his livelihood, and supporting the livelihood of other farmers.

Importance of Farmers' Right to Sell Seed

The pivotal importance of the farmer having the right to

sell (not save nor exchange, but sell) seed has to be seen in the context of seed production in India. In India, the farming community is the largest seed producer, providing about 87% of the country's annual requirement of over 60 lakh tons. If the farmer were to be denied the right to sell, it would not only result in a substantial loss of income for him but far more importantly, such a step would displace the farming community as the country's major seed provider.

Weak Farmers' Rights, including denial of right to sell seed, will allow seed corporations to dominate the seed market since farmers will be denied the right to function as seed producers. The space vacated by them will be taken by the seed industries since public sector institutions have been so weakened by budget cuts that they could not compete. The seed industry would then become the dominant source of seed.

Strong Farmers' Rights, allowing the farmer to continue to be a significant supplier of seed, make the farming community a viable competitor and an effective deterrent to the take over of the seed market by the corporate sector. Control over seed production is central to food security which is in the forefront of national security.

Other Kinds of farmers' Rights in the Bill

Contribution of rural/local communities: Apart from the right to sell non-branded seed of protected varieties, the rights of farmers and local communities are protected in other ways too. There are provisions for acknowledging the role of rural communities as contributors of landraces and farmers' varieties in the breeding of new plant varieties. Breeders wanting to use farmers' varieties for creating EDVs cannot do so without the express permission of the farmers involved in the conservation of such varieties.

Anyone is entitled to register a community's claim and have it duly recorded at a notified centre. This intervention enables the registration of farmer varieties even if the farmers themselves cannot do this due to illiteracy or lack of awareness. If the claim on behalf of the community is found to be genuine, a procedure is initiated for benefit sharing so that a share of profits made from the use of a farmer variety in a new variety goes into a National Gene Fund.

Disclosure: Other details supportive of the rights of farmers are the explicit and detailed disclosure requirements in the passport data required at the time of applying for a Breeders' certificate. Concealment in the passport data will result in the Breeders' certificate being cancelled. GURT (terminator) forbidden: Breeders will have to submit an affidavit that the variety does not contain a Gene Use Restricting Technology (GURT) or terminator technology.

Protection against innocent infringement: The draft legislation has also attempted to address a concern voiced by several quarters, that when the new system of Plant Breeders' rights is imposed for the first time, there will probably be many cases of unknowing infringement of breeders' rights. Section 43 specifies that the farmer cannot be prosecuted for infringement of rights specified in the Act if he can prove in court that he was unaware of the existence of such a right.

Exemption from fees: Further, protecting farmers from the new set of provisions being put in place, the Bill stipulates that if farmers wish to examine documents and papers or receive copies of rules and decisions made by the various authorities, they will be exempt from paying any fees. Such fees will be payable by all other people.

Clauses that need amendment

Benefit sharing: The use of farmer varieties to breed new varieties will have to be paid for. Revenue will flow into a National Gene Fund. Despite its good intentions of protecting the interests of the farming community, the formulation of this section [46 (2) d] is likely to create problems in implementation because the drafting is poor. The Gene Fund should be the recipient of all revenues payable to the farming community under various heads. Farming communities should collectively, rather than individually, access this money, except in clear cases where an identifiable farmer's variety has been used. Farmers should have the right to decide how this money that they have earned will be spent. The use of the money should not be restricted to conservation or for maintaining ex situ collections.

The method for fixing and realising benefit sharing should be made simpler and easier to implement. One approach to fixing benefit sharing could be a system of lump-sum payments, based for example on (projected) volume of seed sale.

Protection against bad seed: In providing a liability clause in the section on Farmers' Rights, the farmer in principle is protected against the supply of spurious and/ or poor quality seed leading to crop failures. At present there is too much left to the discretion of the Plant Variety Authority which will fix the compensation. This could lead to arbitrary decisions and should be amended. If it is proven that the breeder has made false claims and the farmer has suffered a crop failure, then compensation should be awarded amounting to at least twice the projected harvest value of the crop. Compensation should be large enough to be a deterrent. In addition, a jail term should be provided if the breeder repeats the offence.

Protection of public interest

The new Act includes public interest clauses, like exclusion of certain varieties from protection and the grant of Compulsory Licensing. To secure public interest, certain varieties may not be registered if it is felt that prevention of commercial exploitation of such variety is necessary to 'protect order or public morality or human, animal and plant life and health or to avoid serious prejudice to the environment'.

Compulsory license

The Act provides for the granting of compulsory license to a party other than the holder of the Breeders' Certificate if it is shown that the reasonable requirements of the public for seeds have not been satisfied or that the seed of the variety is not available to the public at a reasonable price. The breeder is entitled to file an opposition but should the charge be valid, the breeder may be ordered by the Authority to grant a compulsory license under certain terms and conditions including the payment of a reasonable license fee. A compulsory license however will not be awarded if the breeder can demonstrate reasonable grounds for his inability to produce the seed.

Is UPOV appropriate for developing countries?

The interests of developing countries are not served by UPOV, which is completely insensitive to their needs. In all fairness, UPOV was not created for developing countries and therefore does not address itself to their concerns. In understanding the UPOV system, it is crucial to understand that right from 1961, even when it was more flexible than it is today, UPOV granted only one right, the right to the Plant Breeder. There was never any concept of Farmers' Rights. What was granted at best, was an exemption to farmers and researchers, from the otherwise exclusive rights granted to the breeder.

The UPOV is a corporate-backed organisation based in Geneva. It was set-up in 1961 at a time when investments in agricultural research in Western nations were diminishing in the public sector research establishment and moving into private hands. This was a logical step in industrialised economies where the percentage of people engaged in agriculture was reduced to about 2 to 5% of the total population. As the corporate sector made private investments, it sought returns on these investments and set-up UPOV to protect the interests of breeders who were increasingly corporate breeders.

UPOV started as a flexible system, which, apart from granting breeders' rights was not particularly concerned about restricting the exemptions it provided to the other players, namely farmers and other researchers. This began to change as the corporate breeders consolidated their hold on the plant-breeding and seed-producing industry. Amendments to the UPOV convention were brought in 1972, 1978 and 1991.

All these amendments had one goal, to further strengthen the hold of the breeder and reduce any exemptions that were granted in early versions of the convention. The valid treaty of today is the 1991 treaty, which has almost exclusive rights of breeders and no exemptions for farmers or researchers. In fact UPOV has moved to accept the patents system now so that it is not only a platform for breeders' rights but also for patents on plant varieties.

In light of all this and the path-breaking Indian legislation, the recommendation made by the Agriculture Ministry to join UPOV, has stunned the international community and those who have fought hard to get a balanced legislation in India.

The Indian legislation is incompatible with UPOV in the farmers' rights it provides and the strong public interest clauses it contains, which UPOV does not support. It seems likely that the main casualty of UPOV membership will be the issue of farmers' rights. The 1978 version grants only paltry exemptions to farmers and the 1991 revision of the UPOV Convention has further restricted the scope even of exemptions to farmers that can be granted under national law and the trend is for further restrictions.

Any legislation must be 'within reasonable limits and safeguarding the legitimate interests of the breeder' (Article 15.2 UPOV 1991 Convention), The term 'legitimate interests of the breeder' has been widely interpreted to mean compensation or remuneration to the breeder if the farmer even saves seed from his harvest, to plant the next crop.

Senior scientists of the ICAR and bureaucrats of the Agriculture Ministry who have been UPOV's strongest advocates, offer the argument that joining UPOV will allow India to protect its breeders' rights without having to do bilateral agreements. The fact is that Indian varieties for which there is demand outside are essentially sugarcane, spices and some wheat varieties. These are grown in Africa and some Asian countries, none of which are UPOV members. UPOV members like Japan, Korea, USA, EU, Canada, etc. do not grow any Indian varieties. So on the face of it, the decision to join UPOV makes absolutely no sense.

An alternative to UPOV

The Gene Campaign, along with other civil society organisations, believes that UPOV does not have conditions favourable to developing countries6. It is absolutely necessary that developing countries need to craft their own platform, to address their special needs, which are very different to the needs and requirements of, industrialised countries for whom UPOV was developed. Some efforts have already been made to provide an alternative to UPOV. The model law drafted by the Organisation of African Unity (OAU) and the Gene Campaign's Convention of Farmers and Breeders (CoFaB) is being discussed in various forums. CoFaB has been analysed along with UPOV and described by the UNDP as 'a strong and coordinated international proposal which offers developing countries a far better alternative to European legislation, by focusing on the need to protect farmers' rights and food and nutritional security goals of their people'.

CoFaB and the OAU model law reflect the strengths and vulnerabilities of the developing countries. They aim to secure their interests in agriculture and fulfil the food and nutritional security goals of their people. CoFaB and the OAU law seek to achieve inter alia the following goals:

— Supply of reliable, good quality seeds to the small and large farmer.

— Provide for breeders of new varieties to have protection for their varieties in the market, without prejudice to public interest.

— Acknowledge the role of farmers as creators of land races and traditional varieties which form the foundation of agriculture and modern plant breeding.

— Acknowledge the enormous contribution of farmers to the identification, maintenance and refinement of germplasm.

— Provide a mutually acceptable system of access to biological resources, community knowledge, technologies and practices subject to the prior informed consent of the State and the concerned local communities.

— Provide and promote appropriate mechanisms for the enforcement of the rights of local communities, including farming communities, and breeders and the conditions essential for access to biological resources, community knowledge, technologies and practices.

— Ensure that plant genetic resources are utilised in a sustainable and equitable manner so as to guarantee— national food security.

— Maintain genetic diversity in the field.

— Emphasise that the countries of the tropics are germplasm-owning countries and the primary source of agricultural varieties.

10

Advances in Plant Biotechnology

The development of plant biotechnology has been related to the discovery of the growth regulators and their progressive introduction in tissue culture. In 1934, in the United States, white was producing continuous growth of in vitro tomàto roots. In 1957, that 6-furfurylaminopurine, a cytokinin isolated from a DNA hydrolysate, could induce some caulogenesis, but also that the cytokinin/auxin balance could start different morphological programmes. A ratio higher than one unit induced caulogenesis, a ratio inferior at this unit induced roots.

On this basis, a multibillion dollar industry was created. Hundreds of small and large nurseries and biotech labs throughout the world are propagating in vitro more than 1000 different plant species. This micropropagation technique offers not only means for mass propagation, but also plays an important role to conserve elite or rare plants that are threatened with extinction. Moreover, meristem tip culture offers the possibility of virus eradication. These technologies started about 20 to 25 years ago, especially for ornamental plants, and for a few other crops : potato, strawberry, bananas. Propagated clonally, it was very important to have a true to type propagation technique.

On the other hand, it's also possible to use tissue or cell culture to increase genetic variability. Undifferentiated cells obtained from callus, cells or protoplasts culture are produced and submitted to an selective pressure to improve a and submitted to an selective pressure to improve and fix the somaclonal variants.

Production of Virus-free Plants

Virus eradication is dependent on several parameters. But to take advantage of the non uniform and imperfect virus distribution in the host plant body, the size of the excised meristem should be as small as possible. For Stone, only tips between 0.2 and 0.5 mm most frequently produce virus free carnation plants. The explants smaller than 0.2 mm can't survive, and those larger than 0.7 produce plants that still contain mottle virus.

Today, no definitive explanation can be given to understand this virus eradication various explanations have been given : absence of plasmodesm in the meristematic domes, competition between synthesis of nucleoproteins for cellular division and viral replication, inhibitor substances, absence of enzymes preiral replication, inhibitor substances, absence of enzymes present only in the cells of the meristematic zones, and suppression by excision of small meristematic domes. This last proposal could explain why some potato plant showing virus particles in the meristematic domes, could regenerate a virus free plant.

Factors influencing virus elimination

The success in virus elimination depends on the choice of the explants, the virus, temperature,....In general a high temperature (37°C) is effective for virus diseases, but is depressive for viroids.

The introduction of virus inhibitors (ribavirin) into the culture medium can't inactivate the viruses, but it prevents their replication. A combination of cytokinin and ribavirin, or a heat treatment in vitro, are sometimes used with success to inactivate some viruses difficult to eliminate by meristem culture (NRSV, PDV,..).

Micropropagation techniques

During the micropropagation process, the genetic stability. During the micropropagation process, the genetic stability of the new shoots is dependent upon their origin. Axillary shoots issue from pre-existing buds and are normally true to type, indeed the meristematic cells are genetically very stable.

Adventitious shoots, such as somatic embryos, are neoformed buds developed directly on some organs, or indirectly through a callus phase formed on this organ. So, if the mother plant presents a cell mosaic or chimaeric tissues, risks of genetic variation exists. It is similar in the case of an indirect regeneration, when the callus phase is too long.

Propagation by axillary shooting

This technique has proved to be the most applicable and reliable method of in vitro propagation. Axillary shoot growth is stimulated by overcoming apical meristem dominance. Commercial tissue culture laboratories are now able to propagate a large number of herbaceous ornamental species and several woody plants in this way. However, the propagation of Pelargonium, Howea, and a few other horticultural plants are always difficult to propagate by axillary branching.

Propagation by direct or indirect organogenesis

Adventitious shoots could arise directly from the tissue

of explants without callus formation. Several plants of the family gesneriaceae (Saintpaulia, Streptocarpus,...) regenerate directly buds on leaf explants, likewise Lilium rnerate directly buds on leaf explants, likewise Lilium regenerates on scales. However more often, like for Ficus lyrata, adventitious buds appear on callus. While coffee, cocoa trees, and many conifers are produced by somatic embryogenesis developed on callus or cell suspensions.

Constraints of in vitro micropropagation

The establishment of axenic culture could be difficult when the explants are coming from hot and humid countries. They add that microbial contamination in commercial plant tissue culture laboratories is the most important cause of losses. For this reason, they support the introduction of a microbiological production control strategy.

A second important constraint during the initial phase mainly concerns woody plants. A mature tree must be rejuvenated to recover its morphogenetic competence. Franclet described various treatments, micrografts, cascade grafts, cytokinin treatments to reverse the adult phase to juvenile. The micropropagation of Acacia Senegal was only possible after rejuvenation by micrografting (Palmnly possible after rejuvenation by micrografting.

A distinction between chronological, ontogenic and physiological age is needed to understand the rejuvenation concept. This was definitively demonstrated after a complete reversion of Sequoia sempervivens.

Axillary branching

The cost of micropropagated plantlets is also an important limitation of the techniques. In New Zealand, where they produce 2-3 million micropropagated radiata pine per annum, the relative cost of micropropagated planting stock had dropped from 13,8 times the cost of seedlings in 1988 to 6,9 by 1993.

To reduce manpower costs, several improvements have been proposed. The more simple method was in vitro layering developed to clone PVX-free potato plants. The first plantlets placed on the medium in a horizontal position developed axillary shoots. They are harvested by cutting one centimeter above the medium surface, at 3 weeks intervals.

Somatic embryogenesis propagation

For genetically stable species, somatic embryogenesis offers a very fast scaling-up system, especially when it's possible to produce embryos in bioreactors. Unfortunately this production via fermentors was not as simple as first envisaged. And today, only a few model plants are successfully produced by such technology : carrot, celery. Other applications remain at the experimental stage : coffee, oil, palms, conifers, Euphorbia pulcherrima,... and several other horticultural species.

Several bottlenecks limit the use of this interesting technology. One of the main problems is genetic stability. So, despite the clonal nature of nucellar embryos, different morphological anomalies can occur among mango somatic embryos, as it was also observed in Citrus plants derived from nucellar cultures.

Another difficulty is the loss of embryogenic capacity over time, a phenomenon observed with different species. It is als a phenomenon observed with different species. It is also important to note that somatic embryogenic lines of conifers are always originated from immature embryos.

Artificial seeds

Another very interesting possibility of the somatic

embryogenesis technology has been developed during the past fifteen years. They were able to encapsulate somatic embryos by hydrogel coatings (sodium alginate), producing single embryo artifical seeds. To date, some improvements offer the possibility to directly plant the artificial seeds in the greenhouse on special substrates (vermiculite, sans,...). This methodology will provide in the future a good technique to reduce the cost of transplants.

Gene bank

Tissue culture methods offer the opportunity for in vitro collecting, rapid multiplication and distribution of important, elite, or rare plants that are threatened with extinction. The two major in vitro storage strategies are slow growth and cryopreservation.

Dereuddre et al. have provided a very simple technology to freeze encapsulated meristems in dried alginate beads. It works for pear, strawberry, eucalyptus, potato,.... Some international institutions have very large collections of old and current varieties, available for exchange and introduction in crop improvement programmes. The International Potato Center (CIP) in Lima, Peru, has a large world potato collection. Germplasm of sweet potato and cassava is at the International Institute of Tropical Agriculture (I I T A), Ibadan, Nigeria.

The movement of germplasm involves the risks of accidentally introducing plant quarantine pests along with the host plant material. To limit these risks, the plant material should be transferred from one country to another as in vitro cultures through a transit Centre, where it should be indexed. For bananas, in the framework of INIBAP, the transit Centre is the Catholic University of Leuven in Belgium where a very large in vitro germplasm exists.

Increasing genetic variability through Tissue Culture Techniques

Somaclonal variations

The production of plantlets by callus regeneration, cell suspensions, protoplast cultures, could present some deviations with regard to the mother plant. This is a way to increase the genetic variability. Associated with a selective pressure (stress to toxins, pH, salinity, cold,...) it's used to obtain resistant lines. Indeed, after regeneration, plants can express new potentialities rarely obtained another way. Stable and profitable variants are seleanother way. Stable and profitable variants are selected and introduced in breeding programmes.

Haplodiploïdisation

Many species are able to produce haploïds through different in vitro techniques. The oldest was anther culture of androgenesis. To date, the results vary considerably from one species to another. Solanae (datura, tobacco, red pepper, eggplant, petunia), cereals (wheat, barley, rice, triticale, maize) or crucifers (soybean, cabbage) are species easy to regenerate by anther culture. To the contrary, tomatoes, leguminous, or Compositae are recalcitrant.

Ovule culture, or gynogenesis, was a successful technique for Gerbera, beet, courgette. However, the most common technique is pollination with irradiated pollen, in order to induce in vivo parthenogenesis. The oospheres developed in embryos without fertilisation are saved by embryo rescue.

The haploid plantlets (n chromosomes) are tre. The haploid plantlets (n chromosomes) are treated with colchicine to produce fertile homozygous lines, called "double haploid lines". This haplodiploïdisation technique could give immediately new elite genotypes, hybrid parents after in vitro propagation, as asparagus supermales (MM), or useful genetic material to establish gene mapping.

Protoplast culture

Protoplasts are the smallest units able to regenerate a whole plant. Therefore protoplasts cultures can serve to enlarge genetic variability by introducing somaclonal variation. However, the main interest of protoplasts, is their capacity to fuse and to produce hybrids or cybrids, new organisms most often unknown in nature. Indeed, protoplasts allow to transgress the barrier of the botanical species or genus.

Naked protoplasts can accept without rejection external elements : nuclei, cytoplasmic corganelles, liposomes, containing genetic information. The first protoplast fusion application was a cytoplasmic transfer from one genotype to another to induce male sterility (CMS) from mitochondrial origin. These male sterile hybrids are interesting to produce F1 hybrids (Brassica, Cichorium).

Another cytoplasmic transfer concerns the introduction of chloroplastic DNA to induce herbicide (atrazine) resistance, from Solanum nigrun duce herbicide (atrazine) resistance, from Solanum nigrun to tomato for instance. Many intergeneric and interspecific hybridations are possible but not always profitable because these hybrids are too often infertile. This problem does not exist if we can vegetatively propagate the hybrids, as in the case of Citrus rootstocks.

Molecular biology in Horticulture

More and more pathologists are using molecular probes for early detection of diseases. However, the main contributions of modern biotechnology remain probably in the hands of the plant breeders. Indeed, molecular markers are of great use in detecting desirable new genes, or to identify important QTL. This marker-assisted selection is very useful for many vegetable plants. Other applications are DNA - fingerprinting and genetic engineering. This recent technology aims to insert and to express new specific genes into a selected plants.

Marker-assisted Selection

Molecular markers and genome mapping will reach a large extension in the next breeding programmes of tomatoes, peas, cabbages, melons, potatoes,... It will be possible to improve the speed and the incorporation efficiency of desirable new genes by utilizing of closely linked selectable molecular markers genes by utilizing of closely linked selectable molecular markers. The number of backcross will be reduced. Tanksley et al. show that only with 12 markers, one on each tomato chromosome, the composition of the recurrent parent genome is similar to that observed in the absence of selection after the third backcross two years later. The utilization of molecular markers could also identify important quantitative trait loci (QTL).

DNA Fingerprinting

The DNA polymorphism observed by RFLP, AFLP, or RAPD, allows cultivar identification of fruit trees, apple, citrus, or other vegetatively propagated plants. Although some mutants, as coloured fruits, escape these identification techniques, some people would evaluate the genetic uniformity of regenerated plants in this way.

Genetic Engineering

Genetic engineering consists of introducing a foreign gene

into a plant genome to create a new function. Or at the contrary, it could be to reduce or suppress an existing gene function, as in the antisense strategy.

Transgenic Plants with New Agronomic Traits

Two different techniques are commonly used to introduce genetic information (DNA) inside cells, protected by pectocellulosic walls. A particle bombardment process, called biolistic, is used with success for monocots gynmosperm, squash, peas etc., Today, this technique allows the bombardment of meristems and avoids the difficulties of regeneration. But the mediation of Agrobacterium remains till now the most routinely system to transform dicotyledonous plants.

The main agronomic traits introduced in horticultural plants and already commercialised are Bt toxin and herbicide resistance.

Antisense Strategy

Calgene created in 1994 the first commercial transgenic plant, a long shelf life tomato, by the suppression of polygalacturonase activity due to an antisense gene. However, t activity due to an antisense gene. However, this Flavr Savr tomato variety was removed from the trade 3 years later, because of its disease susceptibility and its lack of productivity! Later, other tomato varieties with long storage qualities were obtained by the utilization of an antisense RNA inhibition of ACC synthase or ACC oxydase, two ethylene precursors.

The antisense technique was also used to reduce the lignification of woody plants, by blocking the enzymes involved in the precursor of lignin biosynthesis. Another interesting application wsa the induction of white flowers in petunia and in different other ornamental plants by the suppression of chalcone synthase activity.

Transmission of the New Traits

These traits induced by the transferred DNA are transmitted to the progeny as a dominant Mendelian character. Nevertheless in some cases, transgenic plants with a strong gene expression can generate progenies with only a faint no gene expression. This problem is not clearly elucidate.

Marker, Reporter, Promotor and Expression Genes

To select the transformed cells, some marker genes, very often resistant to antibiotics or to herbicides, are attached to the coding sequence, as a promoter or an expression gene. Those allow the new gene to express in the whole plant or in a specific plant tissue.

— Some reporter genes are used to follow the evolution oe.

— Some reporter genes are used to follow the evolution of the transformed cells. The most frequent reporter genes are the gene "gus" ((- glucoronidase) or the gene lux (luciferase).

Horticultural transgenic crops already released in USA. The Flavr Savr tomato was the first genetically engineered whole food approved for commercial sale in 1994. Four other transgenic tomatoes with a delayed ripening were approved later. In 1995, potatoes with Bt genes and a squash cultivar resistant to two viruses were released. Two years later, another squash cultivar resistant to 3 viruses and a papaya line also resistant to viruses were approved.

Two other horticultural crops are waiting the authorities approval : red hearted cichory (Radicchio) with male sterility and resistance to herbicide, and a tomato with Bt toxin. It there are so few commercial transgenic plants in horticulture, it's probably due to the relatively low acreage of these horticultural crops in comparison with other agricultural crops. In 1997, in the world 5 million ha were planted with glyphosate resistant soybeans released by Monsanto alone.

Nowadays, it's true also that many transgenic plants exist in research laboratories, awaiting authorization for field testing, as they are very interesting model plants for learning physiology : lettuce with less nitrate by increasing nitrate reductase gene activi with less nitrate by increasing nitrate reductase gene activity, a yeast ribonuclease gene to reduce viroid infection on potato, modification of lignin synthesis in plant to produce trees adapted to paper industry, or timber, or biomases production, regulation genes to control tree architecture,....

A general use of in vitro micropropagation in the developing countries is now a reality, as it was predicted by Albert Sasson. African countries for instance are producing potatoe, banana, cassava. Even the production of transgenic plants is no longer a prerogative of the Northern countries. Different experiments are now starting in the South, specially in the international laboratories from CGIAR. Although CGIAR's expenditures on biotechnology for 1997 raise only 24.2 million US$ on an annual budget of 345 million US$ this Consultative Group on International Agricultural Research plays a very important role as contributor to agricultural research for developing countries.

The large international research centres IITA (Ibadan, Nigeria), CIP (Lima, Peru), CIAT (Cali, Columbia),... belong to the CGIAR. Their roles and NGO's are primordial. The genetic transformatheir roles and NGO's are primordial. The genetic transformation of cassava is an excellent example. Its tuberous roots provide food for over 500 million people, mostly small-scale farmers. Unfortunately, till recently this integral plant for food security in developing countries has been recalcitrant to transformation approaches. However, thanks to the participation of international institutes located in cassava growing countries, CIAT and IITA with the help of four development associations from Swiss, UK, The Netherlands and USA, were able to initiate procotols for cassava transformation and regeneration.

Other networks, like the Cassava Biotechnology Network (CBN) previously described, exist. REDBIO, a technical cooperation network on plant biotechnology, is supported by FAO to promote a best use of scarce manpower, equipment and other resources in Latin America.

On the other hand, more and more private laboratories exist in the developing countries which have a high scientific and technical level (China, Singapore, Taiwan, India, Brasil, Mexico, Chile,...). In these countries, European, American or Japanese agrochemical companies are investing. Also these countries will have their chance to develop their own technologies, and to export to the Northern markets.

Meanwhile, genetic engineering techniques are being applied mainly to crops which are important for g applied mainly to crops which are important for the industrialized world, not crops on which the world's hungry depend. Therefore, it is unrealistic when Monsanto writes "the experts said that biotechnological innovation will increase the crop productivity without occupation of new lands, saving tropical forest of best quality and animal habitat". However, the World Bank was promising in a 1997 publication Bioengineering of crops that" transgenic crops could improve food yield by up to 25 per cent in the developing countries and could help to feed an estimated additional three billion people over the next 30 years".

But, not all people are confident in this prospective. To the contrary, some are afraid that the poorest countries, where more than 700 million people are chronically undernourished will miss the biotechnological revolution. Nevertheless, there is some hope to transfer to the developing countries high technology, which don't undermine the environmental and social network.

The GMOs invade progressively all the US lands : 1.6 % in 1996, 15 % in 1997, 50 % in 2000..... It will represent an estimated market of 3,9 million US$ in 2003 ! The largestt agrochemical companies are merging to become giants of agrobusiness such as Novartis, Monsanto, Zeneca etc.,

The application to agriculture of these new technologies certainly opens interesting perspectives, but also raises potential problems. The risk of crop transgene spreading has been demonstrated. A researcher of Clemson University in South Carolina reported "that in a population of wild strawberries growing within 50 meters of a strawberry field, more than 50 % of the wild plants contained marker genes from the cultivated strawberries".

Carbohydrate Metabolism

Several important attributes distinguish plants from other organisms and help explain the singular nature of plant carbohydrate metabolism. These factors result in a unique carbohydrate metabolism that only superficially resembles carbohydrate metabolism in other organisms.

— Plants are autotrophic.

— Plants are photosynthetic.

— Plants are sessile.

— Plants contain plastids.

— Plants have cell walls.

Plants are autotrophic. Almost all plants convert simple nutrients such as carbon dioxide, water, and inorganic ions into all the intermediates required for the biosynthesis of nucleic acids, proteins, lipids, and polysaccharides as well as coenzymes and numerous secondary metabolites. Thus, in addition to fulfilling the energy requirements of the cell, plant carbohydrate metabolism must feed numerous anabolic pathways. This biosynthetic investment affects the regulation of carbohydrate metabolism.

Central role of carbohydrate metabolism in the supply of carbon skeletons for biosynthetic reactions. G6PDH, glucose-6-phosphate dehydrogenase; 6PGDH, 6-phosphogluconate dehydrogenase; PFK, ATP-dependent phosphofructokinase; FBPase, fructose-1,6-bisphosphatase; PFP, pyrophosphate- dependent phosphofructokinase.

Photosynthetic electron transfer generates ATP and energy-rich reductants used to assimilate carbon, nitrogen, and sulfur. Fdx, ferredoxin.

Plants are Photosynthetic

Apart from a very few parasitic species, plants harness light and use its energy to fix and reduce carbon dioxide. The resulting triose phosphates produced can supply carbon to leaf cells or be converted to sucrose for export to other parts of the plant. Other activities associated with the capture of light energy include nitrite and sulfate reduction and ammonia assimilation. In addition, photosynthesis supplies many biosynthetic pathways with reducing equivalents and ATP. Because the processes associated with light absorption are diurnal, plants must cope with major variations in the supply of nutrients during the light and dark periods. This imposes the need for a flexibility in metabolism that is not seen in other organisms.

Plants are Sessile

Terrestrial plants are unable to relocate when faced with physical or chemical stress and must tolerate changes in temperature; fluctuations in the availability of light, water, and nutrients; and herbivory. To ensure survival in unforgiving circumstances, plant metabolism is highly flexible. In addition, plants expend considerable resources in the production of defense compounds.

Plants Contain Plastids

These organ-elles, found only in plants, perform most of the biosynthetic reactions that occur in plant cells. Duplication of some pathways in both cytosol and plastid, and the requirement for coordination of metabolism in all parts of the cell, have resulted in the evolution of several plastid envelope transporters. The properties and functions of these transporters differ among various tissues and plastid types. Besides linking the cytosol and plastid compartments, these transporters also play important roles in metabolic regulation. The segregation of synthesis reactions into plastids presents another problem for the plant cell. Syntheses require an abundant supply of ATP and reducing power in the form of NAD(P)H, both of which must be generated within the plastid. Various mechanisms have evolved to satisfy this need.

Plants have Cell Walls

Cell walls are composed of cellulose, hemicellulose, and lignin. Synthesis of these wall components represents a major drain on carbohydrates, because the pathways for the synthesis of wall components can account for 30% or more of cellular carbohydrate metabolism.

The topic of carbohydrate metabolism is typically discussed in terms of discrete anabolic and catabolic pathways. However, these pathways can be envisioned as a series of "pools" of metabolic intermediates linked by reversible enzyme reactions. Metabolites can be added to or withdrawn from these pools to serve the needs of the various pathways.

Two pools feed plant carbohydrate metabolism. One is composed of hexose phosphates, the other of pentose phosphate pathway intermediates and the triose phosphates glyceraldehyde 3-phosphate and dihydroxyacetone phosphate. The direction of flow through these pools depends on the requirements of the cell. In photosynthetic tissues, carbon flux through pathways also is governed by whether cells are in light or darkness. Metabolite pools in the cytosolic and plastid compartments communicate by way of highly specific carrier proteins.

For example, glycolysis in animals is typically described as the catabolism of either glucose or the storage polysaccharide glycogen to pyruvate. Starch metabolism in plants, however, is compartmentalized in plastids, and the complements of glycolytic enzymes present are different in the cytosol, chloroplasts, and heterotrophic plastids. When viewed as only one of many pathways linking the hexose phosphate and triose phosphate/ pentose phosphate pools, glycolysis is revealed as a set of reactions contributing to an integrated whole plant metabolism rather than a discrete pathway to be memorized and quickly forgotten.

Transporters located in the inner envelope membrane of plastids exchange inorganic phosphate (P i) for carbon metabolites. In the transport processes illustrated here, photosynthetic chloroplasts export C3 intermediates, whereas the non-photosynthetic plastids import a more diverse group of sugars and organic acids. All the transport reactions are reversible, and the direction of transport is dependent on the concentrations of P i and metabolites on each side of the envelope. 3-PGA, 3-phosphoglycerate; 2-PGA, 2-phosphoglycerate.

Mechanism for supplying cofactors to plastids. Cofactors such as NAD(P)H or ATP are required for biosynthetic pathways. (A) Cofactors can be generated within the plastid by catabolic pathways located in the organelle, or by light energy in the case of the chloroplast. (B) A simple shuttle system for the transfer of reducing equivalents into a plastid. Cofactors may be imported from the cytosol either by direct transport (e.g., exchange of ADP for ATP) or via shuttles, as shown here.

A metabolic pool, showing four metabolites at equilibrium. Flow through this pool will be dictated by the addition of specific metabolites (A), or withdrawal of intermediates (B and C). In many cases, the metabolic status of the cell will determine whether a particular compound (D) enters or leaves the pool.

The use and synthesis of cofactors (NADH, NADPH, and ATP) in photosynthetic and nonphotosynthetic plastids. The contribution of photosynthetic electron transfer and photophosphorylation to the supply of cofactors in chloroplasts has not been included. 1,3-BisPGA, 1,3-bisphosphoglycerate.

Carbohydrate Metabolism and Activity

The primary nutrients are used by the organism in a variety of ways. Their breakdown in the digestion results in metabolites that can be utilized for further metabolic processes in the organism. They can be used for the formation of large compounds that contribute to the structure of the organism (like the carbohydrates in the cell wall of plants, proteins in collagen, or the lipids in membranes). The metabolites can also be converted to substances that regulate the organism and its functions (as for instance in immune recognition, neurotransmission, or as enzymes). And they can be oxidized further to provide energy for the above processes and bio-mechanical and active transport needs through their ultimate breakdown, as happens for instance with glucose.

Carbohydrates, proteins, and lipids can each be used for a variety of different functions; within limits they can also be converted one into the other with the help of intermediates in the citric acid cycle, which interconnects their anabolic/catabolic cycles.

Carbohydrates and Water

Water stands at the beginning of the synthesis of carbohydrates in photosynthesis. This is one important process that shows the relationship between carbohydrates and water. Water is oxidized to convert sunlight into biochemical energy for the synthesis of carbohydrates from CO_2:

$$6\ CO_2 + 6\ H_2O + \text{light}\ (\ C_6 H_{12} O_6\ \text{(carbohydrates)} + 6\ O_2$$

Animals and humans breathe in the O_2 and use it to break down carbohydrates to CO_2 and water in the citric acid cycle and oxidative phosphorylation. This oxidative catabolic process, together with the reductive anabolic process of photosynthesis, forms a cycle in nature between plants and the organisms of animals and human beings in which water is alternately oxidized and formed.

The water/carbohydrate connection is also found when sugar is burnt without the addition of oxygen, i.e. made into carbon. This happens when sugar is heated up in a closed container so that no oxygen can enter. The result will be carbon and water, the latter escaping as steam. This phenomenon would imply that the general formula for carbohydrate monomers could be written as Cn (H_2O)n. This is the reason they were originally given the name carbo-hydrate. The general formula for glucose for instance is $C_6 H_{12} O_6$.

Carbohydrates have a relation to water both in their structure and in their metabolic processes. In the formation of polysaccharides, as in forming the peptide bonds in proteins, one water molecule is released from glucose in forming the glycosidic bond with every addition of a glucose molecule.

Functions of Carbohydrate Polymers

Carbohydrate polymers, like all large compounds, hold within their structure potential energy that becomes available at their breakdown as metabolic and other energy for the organism. Many common carbohydrates are polysaccharides. Polysaccharides are polymers of sugars. The most commonly occurring monomer is glucose. Carbohydrates also have structural functions.

Storage function

Glycogen and starch are storage forms for glucose. Glycogen is usually more branched than starch, which allows for the glucose to be cleaved off at more points at the same time. Therefore glycogen, which is the main glucose polymer in higher organisms, can give a greater rate of supply of glucose in demand situations than starch, which exists in plants only, would be able to.

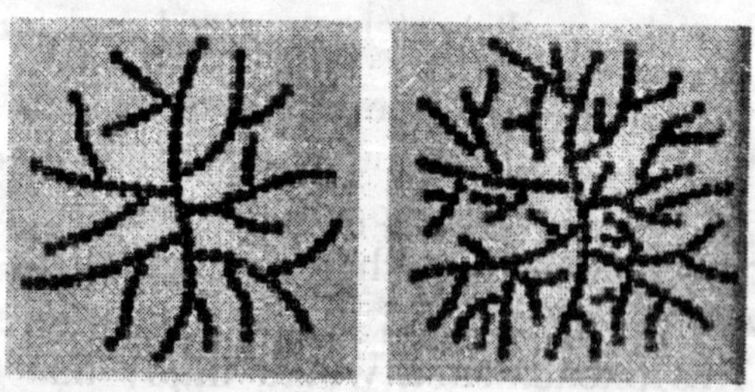

Animals and humans have more glucose quicker at their demand than plants. Cellulose is a structural glucose polymer in plants. Structural carbohydrates in non-plants have amino acids or contain amino acid sequences as monomers to form chitin or peptidoglycans. The breakdown of cellulose does affect the structure of the plant, as does the breakdown of peptidoglycans by antibiotics affect the structure of bacteria, and the breakdown of chitin affects the structure of invertebrates that carry it in their exoskeleton.Plants are the only organisms that use homopolysaccharides to build up their structural components, as well as using starch and glucose for an energy source.

Green Revolution

Asia's 'Green Revolution' achieved increases in crop productivity that were sufficient to lower the proportion of the population suffering from chronic hunger from 40% to 20% while the overall population more than doubled. In addition, by increasing yields on land already in production, hundreds of millions of hectares of tropical forests and other natural environments were saved from conversion to agriculture. Yet, worldwide, roughly 800 million people remain chronically undernourished - consuming less than 2000 calories per day. Even more are malnourished due to diets that are deficient in vitamins and mineral micronutrients.

Many of the most seriously deprived live in rural areas where the Green Revolution technologies have not been adopted, primarily because of inadequate or unreliable rainfall and lack of irrigation . In many developing countries, the majority of the population still produce most of their own food and depend on small-scale farming for their incomes and livelihoods. China and India each have more than 500 million people who depend on small-scale agriculture and Africa has more than 400 million. On the basis of these numbers alone, these countries stand to benefit most from further genetic improvements of crops using conventional plant breeding, as well as breeding supplemented by new biotechnologies that are derived from advances in plant molecular and cellular biology.

Realization of such benefits can already be seen with the adoption of transgenic crops. More than 5.5 million farmers worldwide grew transgenic crops in 2001, including farmers in Argentina, China, South Africa, Mexico, Uruguay, Indonesia and, without government approval, Brazil. Of these, over 75% are small-scale cotton growers, mainly in China, who have readily adopted new cotton varieties that contain transgenes for insect resistance.

The larger developing countries are investing substantial public resources in plant biotechnology research, with China's investment alone now more than US$100 million year and projected to increase significantly . Unlike the USA and Europe, where the public sector has developed many useful transgenic crops that are not being commercialized because of proprietary property and regulatory constraints, the public sector in China is delivering products.

Most of the Bacillus thuringiensis (Bt) cotton varieties and all of the transgenic tobacco and tomato lines commercialized in China were developed by public research institutions. And, it's not just China; promising transgenic lines of around twenty different crops produced by public research institutions in at least ten developing countries are now being officially field tested as they near national regulatory approval. Transgenes have been introduced into many important tropical crops. The majority of the transgenic lines have transgenes for traits, such as virus resistance, that can significantly benefit poor farmers who cannot afford more expensive disease control strategies and currently suffer significant crop losses.

Indeed, public research institutions in countries such as China, India and Brazil, which have both excellent scientific capacity and greater 'freedom-to-operate', are likely to become the primary employers of plant biotechnology to deliver useful new varieties of tropical crops to farmers with limited purchasing power. The private sector is increasingly concentrating on only a handful of major crops and profitable markets. And, owing to proprietary property constraints, public sector institutions in industrialized countries find it increasingly difficult to commercialize products of plant biotechnology without corporate sponsors.

Bt cotton

Transgenic cotton varieties containing insect-resistance genes derived from the insecticidal microbe Bacillus thuringiensis are now being grown commercially in China, South Africa, Mexico, Argentina, Indonesia, and India. Pray et al. have followed the adoption of Bt cotton in China, which began in 1997. By 2001, 3.5 million farmers, growing on average 0.42 hectares, planted 1.5 million hectares of Bt cotton. This equates to roughly 31% of the area planted to cotton in China. The rapid spread of Bt cotton was driven by the farmers' demand for a technology that increases yield, reduces insecticide use and costs, reduces insecticide poisonings and requires less labor.

Initial yield increases were in the 5-10% range and modest increases continued over time, suggesting that farmers are learning to manage Bt varieties better. There is no indication that insect pests are becoming resistant to Bt cotton. The use of insecticides in China has reduced substantially due to the use of Bt cotton. The use of formulated insecticide fell by 20 000 tons in 1999 and 78 000 tons in 2001, the latter being roughly a quarter of all of the insecticide sprayed in China before the adoption of Bt cotton.

Cost savings for farmers are now beginning to push down the price of cotton, so consumers will also benefit. Bt technology is being used increasingly in China as a component of integrated pest management strategies. As documented by Fang et al., much of the Bt cotton grown in China was originally developed by the Beijingbased Biotechnology Research Institute of the Chinese Academy of Agricultural Sciences. This governmentsponsored research institute has obtained patent, plant variety and trademark protection in China for its insectresistant Bt cotton.

The original transgenic lines were sub-licensed to provincial seed companies and the transgenes were backcrossed into more than 22 well-adapted local varieties. Sub-licenses also have been issued to Chinese companies for the commercialization of Chinese Bt cotton technology in Vietnam, Cambodia, Thailand and India. In South Africa, Bt cotton is being grown by both largescale and small-scale farmers, with small holders who farm intensively gaining the most from the technology. The greatest benefit for all farmers is a significant reduction in insecticide costs.

Table 1 summarizes the direct costs and benefits of Bt-cotton production for those farming 1-3 hectares of rain-fed cotton in the Makhathini Flats region of Northern Kwa-Zulu Natal during the 1999-2000 growing season. These farmers, half of whom are women, received a 77% higher return from Bt cotton than from conventional varieties and also benefited through a significant reduction in the number of necessary insecticide applications, which are labor intensive and dangerous for smallholders.

Table 1. Costs and benefits of Bt cotton production on the Makhathini Flats, South Africa 1999 – 2000.

	Conventional cotton	Bt cotton
Yield (kg ha^{-1})	261	471
Value of output (Rand [R] ha^{-1})	568	905
Seed cost (R ha^{-1})	91	197
Insecticide cost (R ha^{-1})	116	72
Gross margin (R ha^{-1})	361	638

In the states of Coahuila and Durango in north central Mexico, Traxler et al. credit Bt cotton with solving a pest infestation problem caused largely by pink bollworm and tobacco budworm that threatened an industry of small-holder (2-10 hectares) and medium-sized (30-120 hectares) producers. The Bt cotton was deployed as part of the government's integrated cotton pest management program. Bt varieties were grown on 96% of the area sown to cotton and reduced insecticide use by 80%.

Pest levels declined to a new low and reached an equilibrium with beneficial insects. Farmers in this region are now reluctant to use insecticides for fear of upsetting the new equilibrium. For 1997 and 1998 combined, an estimated economic surplus of US$6 million was generated by Bt cotton in Coahuila and Durango, with 86% accrued to farmers and 14% to seed suppliers.

Bt-cotton varieties, provided through Monsanto alliances, were recently introduced to smallholders in India and Indonesia. In India, cotton is responsible for more than half of total insecticide use, yet yield losses from insects are still 50-60%. On-farm field trials of Bt cotton in years with high bollworm pressure gave gains of 80% over yields provided by conventional varieties. Indian farmers are naturally anxious to access Bt-cotton seed, and the technology is expected to spread rapidly and cover 25%

of the Indian cotton area by 2005 (M Qaim, personal communication).

Insect-resistant Bt Cereals

Bt maize is being grown commercially in Argentina and South Africa as well as in North America and Europe. In South Africa, Bt-maize production includes Bt-whitemaize grown by smallholders for their own consumption. Bt maize has been field tested and is nearing approval in China, Brazil, Egypt and the Philippines. A publicly produced Bt maize is soon to be field tested in Kenya. Field and greenhouse tests of Bt rice produced by public institutions have demonstrated the effectiveness of Bt technology in controlling rice pests in China, India, and Pakistan.

Unlike chemical insecticides, Bt rice effectively controls Lepidopteran pests without inducing the emergence of other rice pests, such as brown plant hopper. As a component of integrated pest management systems, Bt rice has the potential to increase yields and greatly reduce insecticide use in Asia.

Disease Resistance

Several field tests of transgenic crops containing genes for bacterial and fungal disease resistance are under way in developing countries and results have been promising so far. However, none of these crops has yet been commercialized. Rather, the first biotechnology-derived disease resistant lines to be commercialized have resulted from pyramiding natural resistance genes via markerassisted selection (MAS).

In January 2002, the government of Indonesia released two new rice varieties, 'Angke' and 'Conde', which were derived by disease resistance breeding augmented with polymerase chain reaction (PCR)-based MAS to pyramid bacterial blight resistance genes into commercially adapted varieties. These new varieties are the product of more than ten years of international collaborative research efforts that led to a better understanding of population genetics and genome structure of the rice pathogen 'bacterial blight' (Xanthomonas oryzae) and to an increasing inventory of bacterial blight resistance genes from the genomes of Oryza sativa L. and near relatives such as Oryza minuta.

Figure 1 illustrates the selective addition of a resistance gene, Xa5, into the IR64 genetic background to augment the Xa4 resistance gene already present in the IR64 breeding line. Similar work, also focusing on pyramiding bacterial blight resistance genes using MAS, has advanced to field trials in northern India, eastern India and in China's hybrid rice breeding program. In Africa, Cassava mosaic disease (CMD) is the most widespread and damaging disease of cassava. Epidemics ofCMD can reduce yields by 100%, and losses of 20-90% are common throughout Africa.

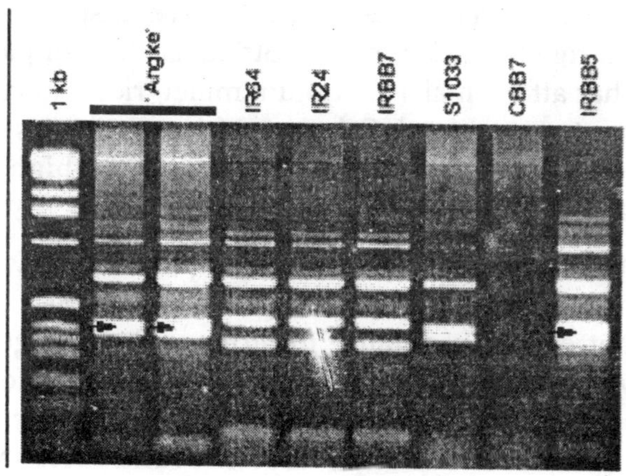

Figure 1.

Farmers cannot afford to use insecticides to control the white fly vector ofCMD, so host plant resistance is the best means of control. A good source of resistance was first detected in third generation backcross progeny from an interspecific cross between cultivated cassava and the wild relative Manhiot glaziovii. Recently, a simple sequence repeat (SSR) marker and a restriction fragment length polymorphism (RFLP) marker linked to a novel dominant gene that confers resistance to CMD were identified in cassava using bulk segregant analysis.

The SSR and RFLP markers were calculated to be 8 cM and 9 cM, respectively, from the gene. The gene was detected in the Nigerian variety 'TME3', and was shown to be qualitative in nature and stable across environments. This result should enable the marker-assisted selection of CMD-resistant cassava genotypes in Africa and in Latin America. CMD has not yet been noted in Latin America, but the vector has recently been found and local varieties are highly susceptible.

Striga Control

Striga, or 'witchweed', compromises several species (most notably Striga hermonthica and Striga asiatica) of parasitic weeds that attack maize, sorghum,millet, rice and cowpea throughout most of sub-Saharan Africa. Estimates of the economic losses caused by Striga reach as high US$7 billion annually.

Maize is particularly sensitive to parasitism by Striga but, unfortunately, researchers have not yet found significant genetic resistance to Striga within the genome of cultivated maize. At present, genetic research is focused on introgressing resistance or tolerance traits that are found in the relatives ofmaize, teosinte and Tripsacum. Meanwhile, a promising seed-based strategy has been developed as an alternative means of controlling Striga.

This strategy is based on resistance to the herbicide imazapyr that was first discovered in 1991 through in vitro selection.

Kanampiu et al. bred imazapyr resistance into maize varieties that were adapted to East African conditions. They then coated the seed of these varieties with a magnesium salt of imazapyr before planting them in soil that was infested with viable Striga seed. During the three months after planting, almost no Striga parasitized plants grew from seed treated with 0.3 mg imazapyr. The few Striga plants that did emerge did not produce flowers and died soon after emergence. In contrast, untreated maize plots suffered from high Striga emergence. These results have since been confirmed on farmers' fields, where the cost of applying herbicide to the seed was just US$4 ha. The development of this method, combined with agronomic control strategies, may provide effective control of Striga in Africa.

Drought Tolerance

Many of the world's poorest people farm in areas with inadequate or unreliable rainfall. Furthermore, agricultural sources of fresh water are decreasing in quantity and quality throughout the world. This is true not only for cereal grain farmers who depend on rainfall for crop production but increasingly for farmers in irrigated areas, which are falling into the 'poorly' or 'partially' irrigated category. Thus, there is a need to achieve greater drought tolerance while not reducing the yield potential of crops in years when abundant water resources are available. The use of transgenics to provide enhanced drought tolerance is still experimental, but progress is being made.

Research on dehydration tolerance has established the bases of at least four signal transduction pathways that function under conditions of abiotic stress and has identified binding elements that bring about stress-inducible gene expression. Wu et al. expressed several 'candidate genes' for abiotic stress tolerance in rice that clearly increased the accumulation of biomass under water deficit. The obvious next step is to investigate the impact of candidate-gene constructs, which are driven by promoters that are induced by water stress, by measuring the growth and agronomic yield of whole plants on a field scale in well-controlled 'managed stress environments'. Hopefully, this can be done by a consortium of plant molecular biologists, integrative plant biologists and agronomists working together in a country, such as China, that is committed to developing droughttolerant varieties.

Currently many research groups worldwide are attempting to demonstrate the success of MAS in breeding for drought tolerance in various cereal crops. A key challenge faced by these groups is determination of genomic regions (i.e. quantitative trait loci [QTLs]) that enhance performance across varying combinations of water-stress conditions, plant growth stages and environments. Ribaut et al. examined the genetic control of the drought tolerance that has been successfully introduced into maize varieties in southern Africa. They focused on the molecular-genetic dissection of component traits that are associated with this tolerance, and identified QTLs that are associated with components of yield of crops under drought stress.

Rice is a hydrophyte, and the enhancement of the rice root system to better extract available soil water during water-deficit periods is a straightforward target for the improvement of drought tolerance. Several studies have focused on augmenting the root system of the broadly adapted rice variety IR64 for greater root depth derived from the upland variety Azucena. Near isogenic lines and specific QTL markers were produced that show promise for the manipulation of genetic potential for root-system depth. Zhang et al. also identified and tagged QTLs for morphological and physiological traits thought to be relevant to drought tolerance in rice.

Recent approaches for improving drought tolerance in pearl millet have focused on the development of QTL molecular markers for drought tolerance during the vulnerable flowering and grain-filling stages. One QTL, which explained 23% of yield under water deficits, was common across environments and has been integrated into pearl millet breeding programs using markers. In sorghum, drought that occurs after flowering is particularly detrimental to yields and the 'stay green' trait (i.e. lack of green leaf senescence) has been associated with greater drought tolerance. Sanchez et al. review the mapping of 'stay green' QTLs for drought tolerance, reporting that four QTLs are consistently associated with the 'stay green' trait in field experiments and explain 53% of the phenotypic variation.

Human Nutrition

Despite progress with supplementation and fortification programs, there is compelling evidence that persistent deficiencies of iron, zinc, iodine and vitamins remain a major cause of numerous human health problems in developing countries. For example, a recent analysis indicates that 127 million pre-school children still suffer from vitamin A deficiencies, leading to blindness and early death. Now, through advances in plant biotechnology there are new opportunities to complement supplementation by including enhanced human nutrition - along with higher yields, reduced losses and greater tolerance of adverse growing conditions - as an important objective when developing crop varieties. Moreover, an effective system for the dissemination of such crop varieties is already in place and has the potential to provide the 'difficult-to-reach' rural poor with nutritionally enhanced staple foods.

Beyer et al. reported further advances in the development of 'Golden Rice': transgenic lines that are engineered to synthesize provitamin A (β-carotene) in the rice endosperm. Mannose has been used as a selective agent so that the new lines contain no antibiotic resistance genes. β-carotene synthesis was achieved by adding only two genes, daffodil phytoene synthase (psy) and bacterial phytoene desaturase (crtI), with endosperm-specific promoters. These new 'clean' lines are being sent to collaborating breeding programs in Asia where they will be crossed with local varieties that are well adapted to the regions where vitamin A deficiency is still prevalent.

Plants as Storage Pesticides

Many of the plant species concerned have also been used in traditional medicine by local communities and have been collected from the field or specifically cultivated for these purposes. Leaves, roots, twigs and flowers have been admixed, as protectants, with various commodities in different parts of the world, particularly India, China and Africa.

Where plants are used as storage protectants they are almost always applied to control insect pests. This is reflected in the volume of research directed to identifying insecticidal or insect repellent plants and plant extracts. Nevertheless, some work has been undertaken to determine whether plants can control storage fungi. Most workers have investigated the properties of spices as inhibitory agents of mycelial growth of Aspegillus species and of its toxin production. Syzigium aromaticum (cloves) have been found to be particularly effective, often completely inhibiting both fungal growth and toxin production.

Many commercially available spices and herbs, turmeric, basil, marjoram, anise, cumin and coriander, are able to completely inhibit toxin production but only partially inhibit fungal mycelial growth. Aqueous extracts of weeds and medicinal plants have also been shown to inhibit toxin production by A. flavus. These include Ricinus communis, Arnebia nobilis and Nicotiana plumbaginifolia. Other fungi, Fusarium solani f. Phaseoli and Verticillium albo-atrum, have been shown to be susceptible to tannins extracted from bark of various trees, including chestnut and wattle.

More recent studies in Ghana have confirmed these properties of spices. Both Ocimum gratissimum and Syzigium aromaticum were very effective in preventing mould growth. However, there is no evidence that farmers actually use spices or other plants for preventing or controlling mould growth in stored commodities.

There is very little information on the rodentical properties of plants except in terms of the pharmacological and biochemical effects reported in medical studies. Despite the cosmopolitan use of plants as grain protectants knowledge of methods and efficacy of these plants is scanty. Most of the information is derived from anecdotal sources, which provide the minimum of information.

Very few systematic studies have been conducted to determine how farmers utilise plant protectants, the methods employed and their effectiveness. Such information is difficult to acquire, requiring the investigator to have a good understanding of the farming and sociological systems operating in the target communities. The introduction of rapid rural appraisal (RRA) and participatory rural appraisal (PRA) techniques in recent years has facilitated the collection of this type of information and since 1992, some studies have been undertaken.

A survey of plants used as domestic insecticides was conducted over a seven month period in 12 out of 15 districts in forest areas of the Ashanti Region in Central Ghana. Fifteen villages or towns were visited in every district and groups of farmers interviewed in each, almost 500 farmers were interviewed individually. Twenty-six different plant species were found to be used as grain storage protectants, the most common being Chromoleana odorata (Siam weed), Azadirachta indica (neem) and Capsicum annum (chilli pepper). Insecticidal activity against insects other than those of stored grain pests was reported in more than 90 other plant species.

The survey found that while approximately a quarter of the farmers used plant protectants in some form, only a small percentage rely on them entirely to protect their harvest from storage pests. Smoking maize stores was the most common method of control in most districts with the exception of the two major areas producing regular maize surpluses where conventional synthetic insecticides are preferred.

In a shorter PRA survey in northern, semi-arid regions of Ghana only 16 plants were identified as being used as grain protectants. Apart from neem, none of the plants featured in the list of stored-product protectants were used in the Ashanti Region. Two of the plants, Chamaecrista nigricans and C. kirkii (both known locally as "lodel'), said to be the most effective, have not been recorded elsewhere, worldwide, as having this property and, consequently, do not feature in any research programmes devoted to plant protectants of stored products; these particular plants, like many of the others used in Northern Ghana, are weeds and serve no other useful purpose.

In Ghana, there is no real tradition of using plants as insecticides in Malawi. Thirty-one groups of farmers, almost a thousand people, were interviewed in a series of surveys conducted in the Northern and Central Regions. Poor quality tobacco leaves, those unfit for sale or sweepings, are used by farmers to protect stored grain. Only in one village was the dried powdered root of a shrub, Dolichos kilimandscharicus, used as a protectant of shelled maize. No other use of plants as storage protectants was recorded though a few others were used to protect growing cereal crops and vegetables and as fish poisons.

It is clear that there is a dearth of information concerning actual use of plants by farmers. There are, without doubt, very many plants used as grain protectants by rural communities, which have yet to be identified. If local production of plant protectants is to be encouraged then it is essential that farm practices are recorded and more information acquired. However, it may prove to be impractical to record treatment application rates and methods because these may vary considerably.

Very little information has been acquired which describes how farmers apply plant protectants; this applies to the three studies mentioned above. This is because farmers are either unable to describe the procedures with sufficient accuracy or their accounts vary considerably from one farmer to another. Thus research programmes which pursue optimal methods of using plant protectants on grain must strive to develop the most cost-effective procedures for application as well as identifying active components.

Various methodologies have been used to determine the effectiveness of plant materials and their extracts. Almost all trials were laboratory-based and of short duration and therefore do not necessarily reflect responses which would be observed under real farm conditions; laboratory experiments cannot account for variables such as fluctuating ambient climatic conditions, the effects of store design and structure and the continuos disturbance of grain by livestock, a multitude of insect pest species and the family itself. Although laboratory investigations can provide useful indicators regarding efficacy of plant products it is essential that trials are also carried out in the field, on farms. Unfortunately, there a very few examples of on-farm or simulated field trials so this is an area of research which must be facilitated and resourced as a priority.

A few trials have been undertaken at research stations in developing countries and these partially simulate on-farm conditions. These trials have mostly utilised plants which have shown considerable promise in the laboratory. Some examples are as follows: Application of Azadirachta indica (neem) seed extract at 8 percent by weight to wheat in jute bags in the Sind, Pakistan was considered to be as effective as 5 mg kg^{-1} pirimiphos-methyl, reducing insect populations after six months storage by 80 percent; calculations demonstrated positive benefits over costs, with ratios of 4.7-7.4. In Nepal, dried powdered rhizome of Acorus calamus (sweetflag) has been found to be effective in almost eliminating insect damage when applied to maize cobs in traditional storage barns.

Attempts to control Prostephanus truncatus, the Larger Grain Borer, using extracts of neem leaves were found to be relatively unsuccessful in simulated storage experiments in both Tanzania and Ghana. Assessment of on-farm use by farmers themselves, using participatory research techniques is lacking. The methods of examination of investigations undertaken in the laboratory to date can be classified into two groups: direct admixture of plant material or utilisation of a solvent extract or essential oil.

Trials of plant material admixed with a commodity, usually directly assess contact toxicity to adult insects, number of eggs laid, number of eggs which hatch or the percentage of first filial (F1) adult emergence. Alternatively, indirect measurements are made by recording the loss in weight due to insect feeding damage, or the number of damaged grains. Assessment of adult mortality may be observed within 3 to 24 hours of treatment, or may be delayed to 7 or 15 days after treatment.

Extracts of plant material rely on the solubility of the active components. Various solvents are utilised, the most commonly used being chloroform, petroleum ether, hexane, methanol, ethanol or acetone. Unfortunately, few workers in this field have reported on attempts to isolate and identify active compounds within the extract. Moreover, the weight or yield of the active component, from a particular extraction technique, is rarely given. Therefore, its concentration on final application to the commodity is not known, although serial dilutions of extract are frequently subjected to insect bioassay.

Essential and vegetable oils may be extracted from appropriate plant species in the laboratory or alternatively commercial preparations may be examined. Vegetable oils are applied at 5-10 ml kg^{-1} commodity although higher values, in excess of 50 ml kg^{-1}, have been recorded. Extracts have been applied to filter paper in various forms of insect bioassay: repellency trials, antifeedant tests and vapour/ fumigation trials.

Antifeedant Tests

In antifeedant tests, extract is applied to wafer discs, filter paper or to paper packing material. The number of holes produced by boring insects are then counted per unit time, usually for a seven day exposure period.

Vapour or Fumigant Toxicity Assessment

For assessments of vapour or fumigant toxicity of essential oils the extract is usually applied to filter paper, left to dry and then suspended in a glass fumigation chamber. Insects are introduced into the chamber and adult mortality, at a given concentration, is then recorded within a defined period of time.

Identification of Active Ingredient

Various solvents have been used in extraction techniques. However, when the plant extracts are tested, few workers have attempted to isolate the insecticidal or non-insecticidal constituents present. Extensive cross-referencing has been necessary to provide an indication of the components or constituents, which may be present in these extracts. Many of the plants which have been tested are found across a broad geographical range, however, only in a few instances has a particular plant species been examined in more than one country. It is well established in plant chemistry that differing climatic, soil and seasonal conditions can affect the type and quantity of the components isolated from extracts.

For example, Acorus calamus is widely distributed in ponds and lakes in Asia, North America and Europe. The rhizomes of the plant contains ß-asarone, a compound which shows toxic and sterilising effects against insects. Three caryotypes of A. calamus have been found: the tetraploid caryotype (4n=48) found in India, East Asia and Japan contains 70-96 percent ß-asarone whilst the diploid caryotype (2n=24) found in North America and the triploid caryotype (3n=36) present in Central Europe contain less than 15 percent of the active ingredient. Researchers in the future will need to take full account of these strain variations when evaluating the potential of particular plant species to avoid utilising plant material from sources with low insecticidal activity.

Bibliography

Clark DG, —Applications of plant biotechnology to ornamental crops" *In: Handbook of Plant Biotechnology - Volume 2: Applications of Plant Biotechnology in Agriculture, the Pharmaceutical Industry and Other Industries.* John Wiley & Sons Press, London, pp 863-879, 2004

De Keukeleire P, De Schepper S, Gielis J, Gerats T: "A PCR-based assay to detect *hAT*-like transposon sequences in plants" *Chromosome Research,* 12 (2): 117-123, 2004

De Schepper S, Debergh P, Van Bockstaele E, De Loose M, Gerats A, Depicker A, "Genetic and epigenetic aspects of somaclonal variation: flower colour bud sports in azalea, a case study", *South African Journal of Botany,* 69 (2): 117-128, 2003

Frossard E, Bucher M, Mächler F, Mozafar A, Hurrell RF, Potential for increasing the content and bioavailability of Fe, Zn and Ca in plants for human nutrition, *Journal of the Science of Food and Agriculture,* 80: 861-879, 2000

Gubrium EK, Clark DG, Barrett JE, Nell TA, "Horticultural performance of transgenic ethylene insensitive petunias" *Journal of the American Society for Horticultural Science,* 125: 277-281, 2000

Harbord RM, Napoli CA, Robbins TP, Segregation distortion of T-DNA markers linked to the self-incompatibility locus in *Petunia hybrida. Genetics,* 154: 1323-133, 2002

Karandashov V, Nagy R, Amrhein N, Bucher M, "Evolutionary conservation of phosphate transport in the arbuscular mycorrhizal symbiosis", *Proceedings of the National Academy of Sciences USA,* 101: 6285-6290, 2004

Karandashov V, Bucher M, "Symbiotic phosphate transport in arbuscular mycorrhizas" *Trends in Plant Science*, 10: 22-29, 2005

Nagy R, Karandashov V, Chague V, Kalinkevich K, Tamasloukht M, Xu G, Jakobsen I, Levy AA, Amrhein N, Bucher M, "The characterization of novel mycorrhiza-specific phosphate transporters from *Lycopersicon esculentum* and *Solanum tuberosum* uncovers functional redundancy in symbiotic phosphate transport in solanaceous species" *Plant Journal, in press,* 2005

Peters JL, Cnops G, Neyt P, Zethof J, Cornelis K, Van Lijsebettens M, Gerats T, "An AFLP-based genome-wide mapping strategy", *Theoretical and Applied Genetics*, 108 (2): 321-327, 2004

Qiao H,Wang F, Zhao L, Zhou J, Lai Z, Zhang Y, Robbins TP, Xue Y, The F-box protein AhSLF-S2 controls the pollen function of S-RNase-based self-incompatibility, *The Plant Cell*, 16: 1-16, 2004

Rausch C, Daram P, Brunner S, Amrhein N, Bucher M, "Phosphorus transport mechanisms in vascular plants: supply meets complementarity", *In: Developments in Plant and Soil Sciences*. Horst WJ et al. (eds.), Kluwer Academic Publishers, Dordrecht, pp 18-19, 2001

Shibuya K, Barry KG, Ciardi JA, Loucas HM, Underwood BA, Nourizadeh S, Ecker JR, Klee HJ, Clark DG, The central role of PhEIN2 in ethylene responses throughout plant development in petunia. *Plant Physiology*, 136: 2900-2912, 2004

Ten Hoopen R, Robbins TP, Fransz PF, Montijn BM, Oud O, Gerats AGM, Nanninga N, "Localisation of T-DNA insertions in *Petunia hybrida* by Fluorescent In-Situ Hybridisation": physical evidence for a suppression of recombination". *The Plant Cell* 1996.

Zimmermann P, Zardi G, Lehmann M, Zeder C, Amrhein N, Frossard E, Bucher M, "Engineering the root-soil interface via targeted expression of a synthetic phytase gene in root hairs". *Plant Biotechnology Journal*, 1: 353-360, 2003

Index